VOLUME ONE HUNDRED AND FIFTY TWO

PROGRESS IN MOLECULAR BIOLOGY AND TRANSLATIONAL SCIENCE

CRISPR in Animals and Animal Models

VOLUME ONE HUNDRED AND FIFTY TWO

PROGRESS IN MOLECULAR BIOLOGY AND TRANSLATIONAL SCIENCE

CRISPR in Animals and Animal Models

Edited by

Raúl Torres-Ruiz
Molecular Cytogenetics and Genome Engineering Group, Centro Nacional Investigaciones Oncológicas (CNIO), Madrid, Spain

Sandra Rodriguez-Perales
Molecular Cytogenetics and Genome Engineering Group, Centro Nacional Investigaciones Oncológicas (CNIO), Madrid, Spain

ELSEVIER

ACADEMIC PRESS
An imprint of Elsevier

Academic Press is an imprint of Elsevier
125 London Wall, London EC2Y 5AS, United Kingdom
525 B Street, Suite 1800, San Diego, CA 92101-4495, United States
50 Hampshire Street, 5th Floor, Cambridge, MA 02139, United States
The Boulevard, Langford Lane, Kidlington, Oxford OX5 1GB, United Kingdom

First edition 2017

ISBN: 978-0-12-812506-9
ISSN: 1877-1173

For information on all Academic Press publications visit our website at https://www.elsevier.com/

Publisher: Zoe Kruze
Acquisition Editor: Ashlie M. Jackman
Editorial Project Manager: Joanna Collett
Production Project Manager: Vignesh Tamil
Cover Designer: Mark Rogers

Typeset by Thomson Digital

CONTENTS

CONTRIBUTORS

Eric P. Bennett
Departments of Cellular and Molecular Medicine and Odontology, Copenhagen Center for Glycomics, University of Copenhagen, Copenhagen, Denmark

Marc Güell
Biological and Biomedical Sciences, Harvard University, Boston, MA, United States;
Harvard Medical School, Boston, MA, United States;
Pompeu Fabra University, Barcelona, Spain

Wenyuan Han
Archaea Center, University of Copenhagen, Copenhagen Biocenter, Copenhagen, Denmark

Izuho Hatada
Biosignal Genome Resource Center, Institute for Molecular and Cellular Regulation, Gunma University, Maebashi, Gunma, Japan

Takuro Horii
Biosignal Genome Resource Center, Institute for Molecular and Cellular Regulation, Gunma University, Maebashi, Gunma, Japan

Michael Kosicki
Wellcome Trust Sanger Institute, Hinxton, Cambridge, United Kingdom

Flaminia C. Lorenzetti
Departments of Cellular and Molecular Medicine and Odontology, Copenhagen Center for Glycomics, University of Copenhagen, Copenhagen, Denmark

Marta Martinez-Lage
Molecular Cytogenetics and Genome Engineering Group, Centro Nacional Investigaciones Oncológicas (CNIO), Madrid, Spain

Emmanouil Metzakopian
Wellcome Trust Sanger Institute, Hinxton, Cambridge, United Kingdom

Yoshiki Narimatsu
Departments of Cellular and Molecular Medicine and Odontology, Copenhagen Center for Glycomics, University of Copenhagen, Copenhagen, Denmark;
GlycoDisplay Aps, Copenhagen, Denmark

John T. Poirier
Department of Medicine, Memorial Sloan Kettering Cancer Center, New York, NY, United States

Sandeep S. Rajan
Wellcome Trust Sanger Institute, Hinxton, Cambridge, United Kingdom

Juan C. Ramirez
VIVEbioTECH, San Sebastián, Spain

Sandra Rodriguez-Perales
Molecular Cytogenetics and Genome Engineering Group, Centro Nacional Investigaciones Oncológicas (CNIO), Madrid, Spain

Qunxin She
Archaea Center, University of Copenhagen, Copenhagen Biocenter, Copenhagen, Denmark

Ellen Shrock
Biological and Biomedical Sciences, Harvard University, Boston, MA, United States; Harvard Medical School, Boston, MA, United States

Raúl Torres-Ruiz
Molecular Cytogenetics and Genome Engineering Group, Centro Nacional Investigaciones Oncológicas (CNIO), Madrid, Spain

Hans H. Wandall
Departments of Cellular and Molecular Medicine and Odontology, Copenhagen Center for Glycomics, University of Copenhagen, Copenhagen, Denmark

CHAPTER ONE

CRISPR History: Discovery, Characterization, and Prosperity

Wenyuan Han, Qunxin She[1]
Archaea Center, University of Copenhagen, Copenhagen Biocenter, Copenhagen, Denmark
[1] Corresponding author. E-mail address: qunxin@bio.ku.dk

Contents

Abstract

CRISPR research is a very young research field since it was only 10 years ago when the system was found to confer antiviral defense. Nevertheless, there has been an explosion of publications in CRISPR research in the past 5 years. The research was started with the comparative genomics of microbial genomes early this century, which revealed the prevalence of clustered regularly interspaced short palindromic repeats (CRISPR) and CRISPR-associated (Cas) in bacteria and archaea. Series of hypotheses were made based on bioinformatics analyses and tested experimentally within a few years after the CRISPR acronym was coined. These findings have not only led to the discovery of the unique antiviral system and the involved molecular mechanisms, but also to the development of CRISPR technology with various well-developed applications, such as genome editing in all three domains of life. Currently, widespread research efforts in multiple research disciplines have constantly yielded new insights

Progress in Molecular Biology and Translational Science, Volume 152
ISSN 1877-1173
http://dx.doi.org/10.1016/bs.pmbts.2017.10.001

into molecular mechanisms of CRISPR antiviral immunity, and new applications in scientific research and biomedical applications. Retrospectively, it is worth pointing out that close interdisciplinary interactions have fostered series of discoveries in the CRISPR research and worked as the driving force in the fast developing research field.

1. INTRODUCTION

Clustered regularly interspaced short palindromic repeats (CRISPR) and CRISPR-associated (Cas) system codes for an adaptive immunity in prokaryotes to defend against invasive genetic elements, including viruses and plasmids. The system is composed of two genetic entities: CRISPR loci consisting of spacers and repeats and operons of *cas* genes. Transcription of CRIPSR loci yields a long transcript termed precursor CRISPR RNA (pre-crRNA) that is processed into small crRNAs of single spacer-repeat unit (mature crRNAs), whereas expression of the *cas* gene operons produces Cas proteins. Then, crRNAs and Cas proteins form nucleoprotein complexes that achieve the immunity by specifically recognizing invading genetic elements via sequence complementarity between the crRNA and its corresponding DNA sequence (protospacer) on invading genetic elements and targeting it for destruction.

The prevalence of CRISPR loci was revealed from the first wave of genome sequencing of archaeal and bacterial genomes in late 1990s and early 2000s. Comparative genomics have yielded a series of discoveries on CRISPR-Cas systems, such as the foreign origin of spacers, extraordinarily diverse Cas proteins with conserved RNA binding, helicase and/or nuclease domains. These results led to a series of predictions on functions and molecular mechanisms of this unique immune system. These predictions were then tested experimentally, revealing unprecedented features for a prokaryotic system: the CRISPR-Cas system provides an RNA-guided antiviral immunity, and the immunity is first gained upon exposure to invading genetic elements. Therefore, CRISPR-Cas codes for an adaptive immune system.

The RNA-guided fashion of DNA interference was then demonstrated experimentally. In particular, investigation of a simple CRISPR-Cas system (later classified as Type II) paved the way for developing CRISPR technology: only a single Cas protein, namely Cas9, is required for DNA interference. Furthermore, it was found that invading DNAs are identified by the presence of a short DNA motif in the corresponding protospacer. As a result, any sequence immediately adjacent to such a motif can be reprogrammed as a target site for specific DNA targeting by a CRISPR-Cas system. Subsequently, several research groups have demonstrated that Type II CRISPR-Cas system can be readily applied in genome editing of human cell lines and mouse models, and since then, there was an explosion in the application of the CRISPR-Cas system for genome editing in many different organisms. In the meantime, great leap forward research progresses were also achieved in investigation of molecular mechanisms of different CRISPR-Cas systems. The main events in CRISPR biology research and CRISPR technology development are illustrated in Fig. 1, which are the focused points of the present chapter.

2. DISCOVERY OF CRISPR-CAS SYSTEMS

2.1 Identification of CRISPR Loci and *cas* Genes

CRISPR repeats were first discovered by accident in 1987. In the determination of the gene sequence coding an alkaline phosphatase isozyme responsible for conversion aminopeptidase in *Escherichia coli*, a peculiar repeat sequence was detected downstream of the gene. The repeat sequence contains 5 29-nt repeats of identical sequence that are interspaced by 32-nt unique sequences,[1] which is a part of the 12 repeat locus clustered with the CRISPR-Cas system in *E. coli* (later classified as Type I-E[2]). Subsequently, similar repetitive sequences have been identified in other *E. coli* strain and the closely related enterobacteria, including *Shigella dysenteriae* and *Salmonella enterica.*[3] Similarly, multiple 36-bp direct repeats (DRs) interspersed by unique spacers of 35 to 41-bp were also found in *Mycobacterium tuberculosis.*[4] In fact, the polymorphic feature of DRs in *M. tuberculosis* strains was found to be useful in strain typing.[5,6] This suggested a prevalence of the DNA element in bacteria on one hand and extremely diversity on the other hand, as the repeat-spacer units can be different in very closely related strains.

Archaeal CRISPR repeat was first discovered in 1993. When Mojica *et al.* investigated the effect of salinity on the growth of *Haloferax mediterranei*, a haloarchaeon that only grows under a high salt condition. These researchers identified a long DNA sequence containing regularly spaced repeats in this archaeal genome although the *H. mediterranei* repeat and that of *E. coli* show no sequence similarity.[7] Subsequently, genomic studies on *Sulfolobus* species, a thermophilic acidophile and its genetic elements revealed that such repeats

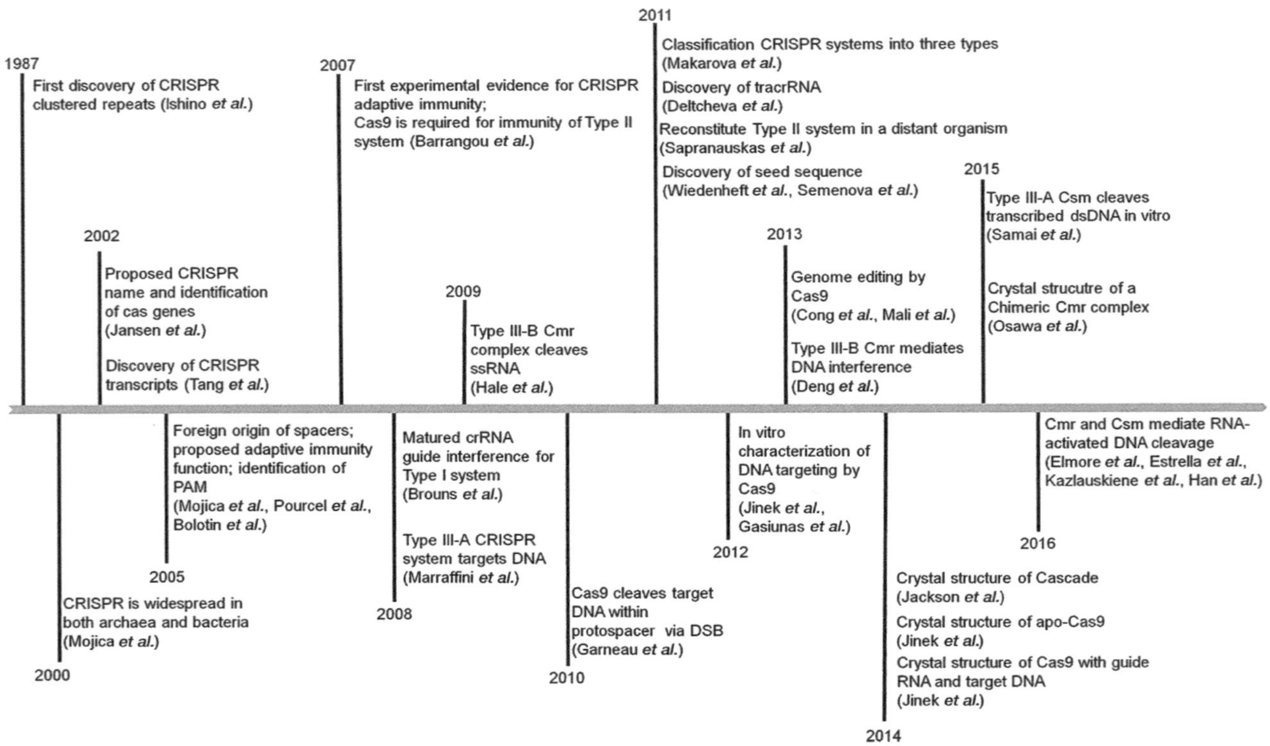

Fig. 1 Time line of the key studies in discovery, characterization, and application of clustered regularly interspaced short palindromic repeats (CRISPR)-Cas system. *Cas*, CRISPR-associated; *Cmr*, Cas module RAMP; *PAM*, protospacer adjacent motif.

are very abundant in the genome of *S. solfataricus*[8] and they are also present on *Sulfolobus* conjugative plasmids, such as pNOB8.[9] In fact, analysis of genome sequences of several archaea and bacteria showed that this unusual repeat structure is widespread in prokaryotes.[10] To date, it is estimated that CRISPRs are present in ca. 40% of bacteria and 90% of archaea.

The function of these repeats remained elusive for about 20 years after its discovery, during which several naming conventions were proposed for this type of repetitive sequences. Those worth mentioning include multiple direct repeats (DRs),[5] short regularly spaced repeats (SRSR),[10] and large clusters of tandem repeat (LCTR).[11] CRISPR was coined by Jansen *et al.* in 2002,[12] which was soon adopted by the research community probably because the acronym better reflects the characteristic structures of these repeats.

In the same paper, Jansen *etal.* also reported that CRISPR loci are closely linked to a group of genes among which several of them are well conserved in CRISPR-containing organisms, but absent from those lacking any CRISPR elements. These genes were termed asCRISPR-associates (*cas*) genes, and the first four *cas* genes (coding for Cas1–4 proteins) were found to be dispersed in gene clusters in an immediate proximity of CRISPR loci.[12] At the same time, Makarova *et al.* also came across these genes from comparative genome analysis of available archaeal and bacterial genomes. These authors found that many of these genes encode nucleases and helicases, suggesting that they could be involved in DNA metabolisms. These proteins were named as RAMPs, originally standing for repair-associated mysterious proteins,[13] but subsequently referring to repeat-associated mysterious proteins to reflect their nature of associating with CRISPR-Cas systems.[14] By 2006, more than 50 families of Cas proteins were identified to constitute several different subtypes of CRISPR-Cas systems.[14,15]

2.2 Prediction of Small RNA-Based Antiviral Functions of CRISPR

2.2.1 Identification of CRISPR-Related Small RNAs

The first evidence suggesting a possible RNA-based mechanism for CRISPR-Cas systems is derived from the archaeal small RNA research in which CRISPR loci were found to be expressed as RNAs of different sizes. Tang *et al.* found that all three CRISPR loci in *Archaeoglobus fulgidus* were expressed to large RNA molecules, which were processed into RNAs of interval sizes corresponding to different numbers of multiple repeat-spacer units and those in the size of single repeat-spacer unit (65–77 nt),[16] and the same was found for CRISPR loci in *S. solfataricus*.[17] Further research on organisms belonging to *Sulfolobus* genus extended the observation not only to CRISPR loci present in *S. acidocaldarius*, but also to a minimal CRISPR locus on the *Sulfolobus* conjugative plasmid pKEF9, in which both mature crRNAs and processing intermediates were identified.[18] Furthermore, CRISPR transcription was found to be controlled by a DNA element called "leader" immediately upstream of each functional CRISPR locus,[18,19] serving as promoter to drive the transcription of the linked CRISPR locus.

2.2.2 Origin of Spacers in CRISPR Loci

Investigation of possible origin of spacers in CRISPR loci attained another breakthrough in revealing the functions of CRISPR. Two pioneer studies conducted independently by Mojica *et al.* and Pourcel *et al.* indicated that extrachromosomal elements, such as bacteriophages and conjugative plasmids, are the main source for spacers, and the existing of specific spacers is remarkably correlated with the resistance to the extrachromosomal elements carrying the sequence of the spacer.[20,21] Further, the second work focused on the highly polymorphic feature of CRISPR loci in *Yersinia pestis* strains, and it was predicted that the new spacers are probably derived from bacteriophages and they are be added from one end of each CRISPR locus, and thus representing a memory of past "genetic aggressions".[21] These studies have, for the first time, linked CRISPR loci to invading genetic elements.

Subsequently, analyses of CRISPR loci present in 24 strains of *Streptococcus thermophilus* and *Streptococcus vestibularis* also revealed the extrachromosomal origin of spacers. Due to the correlation between phage resistance and the number of spacers in CRISPR locus, the authors suggested that transcripts from the CRISPR loci might inhibit phage gene expression by an antisense RNA mechanism.[22] Similar researches were also conducted for the then available genome sequences of archaeal organisms and their genetic elements, and this led to the finding that spacers sequence matches exclusively genomes of genetic elements.[23] By then, the prediction of a small RNA-based CRISPR antiviral defense mechanism was taking shape and the hypothesis is ready to be tested experimentally.

2.3 Experimental Demonstration of the Basic Mechanism of CRISPR-Cas System

In the following 2–3 years, a few researches were conducted from which the antiviral activity by CRISPR-Cas systems was confirmed. These studies have also yielded a three-step pathway of the antiviral activity. These include "adaptation" in which DNA segments from foreign genetic elements are acquired as new spacers in CRISPR loci, "crRNA biogenesis" where CRISPR loci are transcribed and crRNAs are generated by crRNA procession and maturation, "interference" in which crRNAs guide Cas protein to specifically recognize target nuclei acids for destruction.

2.3.1 Adaptation

The idea that prokaryotes could acquire new spacers from extrachromosomal genetic element as a part of immune defense system was experimentally confirmed by Barrangou *et al.* in 2007.[24] They challenged sensitive *S. thermophilus* strains with two phages and found that the survivors acquired phage-derived sequences as new spacers in CRISPR loci next to the old spacers. When exposed to phages again, the survivors exhibited resistance, confirming that spacers that match phage genome provide immunity. Their studies also reveal that deletion of a *cas* gene, *cas7* (renamed as *csn2* later), did not alter existing resistance but resulted in unsuccessful generation of phage-resistant survivors, suggesting that this gene is only involved in acquisition of phage-derived sequences. This study constitutes the first experimental demonstration of antiviral activity, as well as its adaptive nature of the immune system.

2.3.2 crRNA Biogenesis

Although it was shown that CRISPR loci were expressed as large precursor RNAs from which small RNAs of a single repeat-spacer unit were generated, the identity and function of the small RNAs remained elusive. Brouns *et al.* characterized the transcription of the CRISPR locus in *E. coli* K12 in more details and found that small crRNAs contained the last 8 bases of upstream repeat at the 5′-end (5′-handle), the full sequence of each spacer and the 5′-sequence of downstream repeat (3′-handle), termed mature crRNA; and furthermore, crRNAs and recombinant Cas proteins formed in vitro CRISPR-associated complex for antiviral defense (Cascade), which was implicated in antiviral activity.[25] In the meantime, Carte *et al.*[26] showed that *Pyrococcus furiosus* Cas6 was found to specifically recognize and cleave within the repeat region of pre-crRNA transcript.[26] Subsequent investigation of several Cas6 endonucleases encoded in different organisms further confirmed that Cas6 proteins are responsible for generating mature crRNAs.[27]

2.3.3 Interference

In the same article, Brouns *et al.* also showed that the in vivo antiviral activity requires not only *cas* gene coding for components of Cascade, but also the gene coding for the Cas3 nuclease and the spacers complementary either to the template strand or to the coding strand of viral genome. Based on these findings, it was reasoned that the *E. coli* CRISPR immunity is based on targeting viral DNA, not by antisense on mRNA or RNAi, thus representing a completely novel interference activity.[25]

In the meantime, Marraffini and Sontheimer investigated the interference activity by a Type III Csm CRISPR-Cas system in *Staphylococcus epidermidis*.[28] The CRISPR-Cas system belongs to a new subtype designated Csm and putative enzyme for interference was predicted as the largest Cas protein, Csm1/Cas10 that shows no sequence similarity to the Cas3 enzyme of Type I systems.[2] As the *S. epidermidis* Csm system contains a spacer with the perfect match to a region of a nickase (*nes*) gene present in many streptococcal conjugative plasmids. Testing conjugation and transformation of conjugative plasmids in *S. epidermidis* showed that conjugation was prevented in the strain carrying the Csm system; and furthermore, as insertion of a slicing intron into the nickase gene blocked the CRISPR interference, it was reasoned that the interference occurred at the DNA level.[28]

Hale *et al.* took another approach to investigate the antiviral activity by CRISPR-Cas systems of *P. furiosus*. An effector complex named Cas module RAMP [Cmr (repeat-associated mysterious protein)] was purified and characterization of the purified complex revealed that its mature crRNAs contain the same 5′-handle present in the *E. coli* Cascade, but completely lack any 3′-handle sequences.[29] The putative enzyme for interference was Cmr2/Cas10, which shares conserved domains with Csm1, and furthermore, several other subunits from Cmr and Csm also show sequence similarity with each other except the smallest subunit (Cmr5 vs. Csm4) although the Cmr complex only exhibited cleavage activity to ssRNA substrates complementary to crRNA.[29] Nevertheless, it is worth pointing out that both the *S. epidermidis* Csm and the *P. furiosus* Cmr mediate dual DNA/RNA interference as for many other Type III CRSIPR-Cas systems (see Section 3.3).

Impressively, 3 years after the prediction of antiviral activity by CRISPR-Cas systems, CRISPR researches already revealed a striking diversity in effector complexes and their targeting activities. Subsequently, more initiatives were taken in the research community to investigate molecular mechanisms of DNA/RNA interference by different CRISPR-Cas systems identified from bioinformatics analysis and classification.

3. MOLECULAR MECHANISMS OF NUCLEIC ACID INTERFERENCE

In the classification scheme proposed by Makarova *et al.*), three main classes of CRISPR-Cas systems were identified, that is, Type I, II, and III, and their type-specific signature proteins are Cas3, Cas9, and Cas10, respectively.[2] As all tested CRISPR-Cas systems exhibit nucleic acid interference activity, the focus of CRSIPR research at that time was to reveal how each antiviral system could distinguish self versus nonself DNA and selectively target invading genetic elements.

3.1 PAM-Dependent DNA Interference by Type I CRISPR-Cas Systems

As described earlier, the *E. coli* Cascade complex works in concert with Cas3, a helicase/nuclease bifunctional enzyme to exert DNA cleavage.[25] Investigation on how the system could distinguish self vs. non-self DNA led to the identification of the so called protospacer adjacent motif (PAM) element. In fact, the short DNA motif was first identified from the sequence comparison of flanking sequence of protospacers from phages of *S. thermophilus* and *S. vestibularis*,[22] which was then generalized to other CRISPR systems.[30] The presence of PAM sequence was experimentally demonstrated in several antiviral systems, including the I-A system present in *Sulfolobus* species.[31] Further insight into the motif recognition was gained from structural study of the *E. coli* I-E system, where recognition of PAM is independent to 5′-handle of crRNA.[32] A loop region of Cse1, a subunit of Cascade specifically recognizes the PAM sequence, allowing DNA interference to occur.[33] As a PAM sequence is not present in any spacers in CRISPR loci, no spacers would be a target for destruction.

Seed sequence is another element important for CRISPR interference. A detailed study on the complementarity between a crRNA and the corresponding protospacer and mutant derivatives in the *E. coli* I-E system identified a few nucleotides (7 nt) immediately adjacent to PAM that are crucial for DNA interference.[34] Perfect match between crRNA and the seed sequence on protospacers is required for the antiviral activity, whereas multiple mismatches between crRNA and other region of target DNA are tolerated. The mismatch tolerance of the CRISPR immunity allows spacers gained from one virus to be used in DNA interference to related viruses that carry closely related protospacers.[34,35]

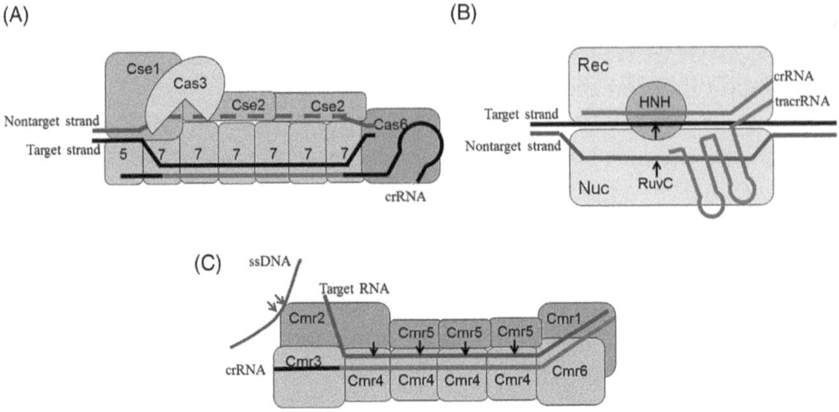

Fig. 2 Models for nucleic acid interference mechanisms of Type I, II, and III CRISPR systems. The representative effectors for each type are shown: (A) Type I-E; (B) Type II; (C) Type III-B. *Cmr*, Cas module RAMP; *Nuc*, nuclease lobe; *Rec*, recognition lobe; *tracrRNA*, *trans*-activating CRISPR RNA.

The structural basis for DNA-targeting by Type I CRISPR-Cas systems has been obtained by investigation of the *E. coli* I-E effector complex (Fig. 2A). The I-E Cascade complex exhibits a seahorse-shaped architecture in which the 3′- and 5′-handles of crRNA are anchored at opposite ends of the complex. The helical backbone, which is consisted of six Cas7 proteins, binds the spacer region of crRNA. Cse1, the largest subunit in the Cascade, is responsible for recognition of PAM and melting dsDNA target at the first two nucleotides next to PAM.[36] This leads to further unwinding of the dsDNA at the target site and to the formation of an R-loop structure between dsDNA target and crRNA. Moreover, the confirmation change of Cse1 may function in recruiting Cas3 nuclease to degrade the nontarget DNA strand in the R-loop structure.[36–38]

3.2 Type II CRISPR-Cas System Requires a Unique tracrRNA for DNA Interference

Cas9, the only Cas protein required for DNA interference in Type II systems was firstly discovered in *S. thermophilus* and *S. vestibularis* by Bolotin *et al.* where it was referred as Cas5.[22] The protein was implicated in the antiviral immunity because it contains a RuvC-like nuclease domain. Indeed, it was found that Cas9 is essential for the antiviral immunity in *S. thermophilus.*[24] Moreover, cleavage products by the

Type II system were analyzed by Garneau *et al.*, and they found that both plasmid DNA and viral DNA were cleaved at three nucleotides upstream of the PAM sequence in vivo, generating blunt ends on the target sequence.[39]

Deltcheva *et al.* investigated crRNA processing in *S. pyogenes* by RNA sequencing technology and identified an additional RNA component for Type II systems. The new species of RNA is abundant and it is transcribed from the upstream of CRISPR array on the opposite strand.[40] The transcript, named as *trans*-activating CRISPR RNA (tracrRNA), carries a 25-nt region showing an almost perfect match to the repeat. It was further shown that the processing of crRNA and tracrRNA is coupled and the process requires the endogenous factor RNase III.

In the meantime, Sapranauskas *et al.* tested the functionality of the *S. thermophilus* Type II system in *E. coli* and found that the CRISPR-Cas system is active in a distantly related organism and it was also demonstrated that Cas9 is the only protein required for interference and that both Cas9 nuclease domains, RuvC and HNH, are involved in DNA interference.[41] Then, two different Type II effector complexes containing Cas9 protein, crRNA and tracrRNA, were found to cleave cleaves dsDNA within the protospacer 3-nt ahead of PAM,[42,43] which occurred exactly the same as previously observed for the in vivo cleavage. Further, these researches revealed that the HNH domain of Cas9 cleaves the strand complementary to crRNA, while the RuvC domain cleaves the opposite strand, suggesting that mutation of one of the two nuclease domain would produce single strand DNA nick. Furthermore, it was shown that fusion of crRNA and tracrRNA to a single-guide RNA (sgRNA) did not impair the efficiency of DNA cleavage by Cas9.[42]

Structure of *S. pyogenes* Cas9 was resolved by several research groups to gain insights into molecular mechanisms of the DNA cleavage. The protein has two structural lobes: recognition lobe (Rec) and nuclease lobe (Nuc) (Fig. 2B).[44] Binding to gRNA leads to reorientation of the two lobes and formation of a channel that can accommodate DNA substrate.[45] The duplex of gRNA and target strand DNA is contacted by an arginine-rich bridge helix (BH) within the Rec lobe.[44] Another arginine-rich motif from the C-terminal domain is responsible for reading PAM on the nontarget strand DNA.[46] PAM recognition and formation of duplex of gRNA and target strand may induce additional conformation changes and promote the cleavage of dsDNA of RuvC domain and HNH domain.[47]

3.3 RNA-Activated DNA Interference by Type III CRISPR-Cas Systems

As described earlier, investigation of the initial two Type III CRISPR-Cas systems revealed that they each carried a distinctive interference activity; whereas the *S. epidermidis* Csm system confers DNA interference in vivo,[28] the *P. furiosus* Cmr effector complex cleaves RNA in vitro.[29] It remained elusive then why the two homologous systems would behave so differently. Interestingly, when Deng *et al.* employed a novel invader plasmid assay to investigate the function of a Cmr module from *Sulfolobus islandicus*, they found the III-B system mediates transcription-dependent DNA interference.[36] In fact, the Cmr-mediated DNA interference is independent of PAM sequence; but has be to be licensed by mismatches between 5′-handle of crRNA and flanking region of protospacer to avoid self-targeting, as reported for the III-A system.[37] Further investigation of the *S. islandicus* III-B system by Peng *et al.* showed that the Cmr system possesses dual DNA and RNA interference activity,[38] suggesting that III-A and III-B CRISPR-Cas systems could be united by the dual interference. Indeed, similar activity was also reported for the *S. epidermidis* Csm system.[48]

The relative ease in reconstitution and purification of Type III CRISPR-Cas effector complexes facilitated their biochemical characterization. The current understanding of DNA and RNA cleavage by Type III systems is summarized by Tamulaitis *et al.*[49] There are three distinctive structural units in Cmr complexes: (a) multiple Cmr4 and Cmr5 subunits form a backbone of two intertwined helical protein filaments; (b) Cmr2 and Cmr3 form a base to anchor the 5′-repeat handle of the crRNA, and (c) Cmr6 and Cmr1 function as a cap to close the seaworm-like complex.[50,51] Csm also contains the three main structural units.[52,53] Both of them cleave complementary ssRNA with 6-nt intervals by the backbone subunit Cmr4/Csm3.[52–54] Furthermore, RNA-activated ssDNA was recently demonstrated for all tested Type III systems with both HD and Palm domains implicated in DNA cleavage (Fig. 2C).[55–58] Further, complementary of 3′-flanking region of target RNA to 5′-handle of crRNA or missing of 3′-flanking region of target RNA abolishes DNA cleavage of Csm and Cmr complexes, suggesting a self-distinguish mechanism at RNA level.[57,58] Clearly, the DNA cleavage is inactivated by target RNA cleavage of Csm3 or Cmr4, which could spatially and temporally regulate the unspecific DNA cleavage activity to avoid self-immunity.[49] The involved mechanisms need to be revealed in future research.

4. DISCOVERY OF ANTI-CRISPR SYSTEM

CRISPR-Cas immunity represents the most powerful weapon of defense in prokaryotes, which is developed in the arms race between prokaryotes and their viruses, but the system is not invincible. Their first enemies were identified by Bondy-Denomy *et al.* in 2013 when investigating a I-F CRISPR-Cas system in *Pseudomonas aeruginosa*.[59] The authors generated 44 different *P. aeruginosa* lysogens, each contains different phage genomes integrated the bacterial chromosome (referred as prophage). These lysogens were then challenged with three phages that would not be replicate in any of these lysogens due to the encoded CRISPR immunity (CRISPR-sensitive phages). Nevertheless, it was found that the CRISPR-sensitive phages robustly replicate in three of the lysogenic strains, and this suggested that the prophages present in the three lysogens must have inactivated the *P. aeruginosa* I-F CRISPR-Cas system. Indeed, screening for putative anti-CRISPR genes (*acr* genes) of phage origin led to the identification of eight genes that confer resistance to CRISPR immunity upon expression in *P. aeruginosa*, allowing CRISPR-sensitive phages to replicate in the bacterium.[59] Subsequently, *acr* genes were found to encode proteins that directly interacting with the I-F Cascade complex or Cas3 nuclease, and these Acr proteins eliminate the CRISPR immunity either by preventing target DNA binding or the recruitment of Cas3 to the Cascade complex.[60] Further work from the same group has led to the identification of more putative anti-CRISPR genes present in *P. aeruginosa* and *Pectobacterium atrosepticum*, and both I-F and/or I-E system could be targeted by the anti-CRISPR mechanism.[61,62]

Although the identified Acr proteins exhibit little sequence similarity, examination of the genomic context of *acr* genes led to the identification of a gene encoding a putative transcription regulator downstream of all *acr* genes, which was referred to anti-CRISPR-associated (Aca) protein.[62] Remarkably, searching the public databases for new Aca proteins resulted in the identification of additional putative *acr* genes in mobile genetic elements (MGEs) that are associated with bacterial II-C CRISPR-Cas system.[63] The function of one such protein, termed AcrIIC, has been demonstrated because it inhibited *Neisseria meningitides* Cas9 activity by directly binding to the Cas protein, and the inhibitory effect was observed in human cells, too. Moreover, Rauch *et al.* developed an approach to discover new Acr proteins by invoking self-targeting.[64] As self-targeting has to be inactivated for cell survival, only those with a functional anti-CRISPR system could survive in the experiments. Four new Acr proteins against Type II-A CRISPR system (ArcIIA) from *Listeria monocytogenes* have been identified, which also inhibit Cas9-mediated genome editing in human cells.[64] The discovery of anti-CRISPR systems reflects the fierce evolutionary arms race between prokaryotes and viruses, which provides an additional layer for manipulating CRISPR-Cas system for biotechnological applications.

5. IDENTIFICATION OF NOVEL CRISPR-CAS SYSTEMS

In the most recent scheme of CRISPR-Cas classification, Makarova *et al.* predicted three novel CRISPR-Cas systems, including Type IV, V, and VI in addition to classic Type I, II, and III systems.[65] Furthermore, to better reflect the diversity of CRISPR-Cas systems, CRISPR class has been introduced as a higher level of classification in which all CRISPR-Cas systems that employ a multisubunit complex for antiviral defense belong to Class 1; whereas those only require a single Cas protein for the activity are termed Class 2 (Table 1).[66]

Representative effector proteins of Type V and VI CRISPR-Cas systems were then characterized, including Cpf1 (CRISPR from *Prevotella* and *Francisella* 1),[67] a system that was previously found to be active.[68] There is a remarkable feature with this Type V system: as for Type II systems, Cpf1 is the only protein required for crRNA processing and target interference, but the effector protein only relies on crRNA for DNA interference and no tracrRNA is required, and that the nuclease cleaves dsDNA target to yield staggered ends of a 5′-4- or 5-nt overhang. As the Cas9 systems, Cpf1 proteins exhibit robust genome editing in human cells.[67]

Several other predicted effector proteins of Class 2 CRISPR-Cas systems were also investigated, and these include C2c1, C2c2, and C2c3.[69] In fact, Cpf1, C2c1, and C2c3 are classified into Type V as they contain distantly related RuvC-like endonuclease domains, while C2c2 proteins contain two predicted HEPN RNase domains and designated as Type VI.[66,69] Type V systems are further classified into subtypes as C2c1 system requires tracrRNA for RNA maturation and mediated DNA interference in 5′-PAM-dependent manner but others do not.

The representative effector of Type VI, C2c2 from *Leptotrichia shahii*, exhibits RNA-guided ribonuclease activity and provides interference against RNA phage.[70] Remarkably, except the specific cleavage of target RNA,

Table 1 An Overview of the Current Classification of CRISPR Systems.

Types	**Class 1: Multisubunit crRNA-Effector Complex**			**Class 2: Single-Subunit crRNA-Effector Complex**				
	I	**III**	**IV**	**II**	**V**			**VI**
Representative effector	Cascade	Csm/Cmr	n.d.	Cas9	Cpf1	C2c1	C2c3	C2c2
crRNA maturation	Cas6	Cas6 and unknown RNase	n.d.	RNase III[a]	Cpf1	n.d.[a]	n.d.[a]	C2c2[a]
Nuclease subunit/domain	Cas3	Csm3/Cmr4; Csm1/Cmr2	n.d.	RuvC, HNH	RuvC, novel Nuc domain			HEPN
crRNA-guided cleavage	dsDNA	ssRNA	n.d.	dsDNA	dsDNA			ssRNA
Collateral cleavage	—	ssDNA	n.d.	—	—			ssRNA

n.d., Not determined.
[a]Existing of tracrRNA.

C2c2 also degrades unspecific RNA when activated by binding to target RNA.[70] The results hint that C2c2 may function in programmed cell death and cell dormancy by degrading host RNA to prevent virus spread.

Most recently, several new Class 2 CRISPR systems were identified by metagenomic analyses from uncultivated bacteria and archaea.[71] These include the first putative archaeal CRISPR/Cas9 systems, and two new bacterial Class 2 systems. One of the archaeal Type II systems contains *cas1*, *cas2*, *ca4*, *cas9* genes, and a hypervariable CRISPR array, suggesting that the system could be active. Attempts were made to test the antiviral activity in *E. coli*; whereas the two bacterial Class II systems are active, the activity of the archaeal Type II system remains to be demonstrated. Strikingly, the bacterial Class 2 systems are only distantly related to any Class II effector proteins previously reported. Moreover, the spacers of only 17–19 nt in length are found to be functional, which are shorter than any spacers present in CRISPR loci known for any other systems. Therefore, it is expected that further exploration of metagenomic data will greatly expand diversity of CRISPR-Cas systems and provide more potential tools for genome editing and other application.[72]

6. BLOOMING OF CRISPR TECHNOLOGY

The development of CRISPR application started with the two pioneer studies and these researches revealed that Cas9-crRNA complexes are active in cleaving dsDNA, indicating that CRISPR-Cas9 systems can be used as a programmed endonuclease for genome editing.[42,43] Indeed, Type II systems (also called CRISPR-Cas9) were successfully applied for genome editing in human cell lines and mouse models.[73,74] These studies lead to a revolution in biological and biomedical researches and biotechnical applications. Now, CRISPR-base cell therapy has been successfully used in clinic and commercial genome-editing service is available. Further, dCas9 (nuclease activity is abolished by mutations ofboth HNH and RuvC domains) with a guide RNA, can function as scaffold for transcription regulation, genome imaging, and epigenetic regulation.[75–80] Due to the simplicity of programming Cas9 system, genome editing and transcription regulation can also be applied on a genome-wide scale, providing a high-throughput approach to assay gene functions.[81–86]

The application of CRISPR system is far more widespread and diverse than summarized here (for reviews, see Refs. 87–90). Nevertheless, a few great challenges remain for any actual application of CRISPR technology in gene therapy, including the occurrence of off-target cleavage and a general lack in methods of tissue-specific delivery. Therefore, joint efforts from multidisciplinary research are highly demanded to further develop CRISPR biotechnology. Conceivably, the advance in the study of CRISPR biology and CRISPR biotechnology will continue to greatly facilitate the CRISPR gene therapy research and other CRISPR-based applications in the upcoming years.

REFERENCES

1. Ishino Y, Shinagawa H, Makino K, Amemura M, Nakata A. Nucleotide sequence of the iap gene, responsible for alkaline phosphatase isozyme conversion in *Escherichia coli*, and identification of the gene product. *J Bacteriol.* 1987;169:5429–5433.
2. Makarova KS, Haft DH, Barrangou R, et al. Evolution and classification of the CRISPR-Cas systems. *Nat Rev Microbiol.* 2011;9:467–477.
3. Nakata A, Amemura M, Makino K. Unusual nucleotide arrangement with repeated sequences in the *Escherichia coli* K-12 chromosome. *J Bacteriol.* 1989;171:3553–3556.
4. Hermans PW, van Soolingen D, Bik EM, de Haas PE, Dale JW, van Embden JD. Insertion element IS987 from *Mycobacterium bovis* BCG is located in a hot-spot integration region for insertion elements in *Mycobacterium tuberculosis* complex strains. *Infect Immun.* 1991;59:2695–2705.
5. Groenen PM, Bunschoten AE, van Soolingen D, van Embden JD. Nature of DNA polymorphism in the direct repeat cluster of *Mycobacterium tuberculosis*; application for strain differentiation by a novel typing method. *Mol Microbiol.* 1993;10:1057–1065.
6. van Embden JD, van Gorkom T, Kremer K, Jansen R, van Der Zeijst BA, Schouls LM. Genetic variation and evolutionary origin of the direct repeat locus of *Mycobacterium tuberculosis* complex bacteria. *J Bacteriol.* 2000;182:2393–2401.
7. Mojica FJ, Juez G, Rodriguez-Valera F. Transcription at different salinities of *Haloferax mediterranei* sequences adjacent to partially modified PstI sites. *Mol Microbiol.* 1993;9:613–621.
8. Charlebois RL, She Q, Sprott DP, Sensen CW, Garrett RA. Sulfolobus genome: from genomics to biology. *Curr Opin Microbiol.* 1998;1:584–588.
9. She Q, Phan H, Garrett RA, Albers SV, Stedman KM, Zillig W. Genetic profile of pNOB8 from *Sulfolobus*: the first conjugative plasmid from an archaeon. *Extremophiles.* 1998;2:417–425.
10. Mojica FJ, Diez-Villasenor C, Soria E, Juez G. Biological significance of a family of regularly spaced repeats in the genomes of Archaea, Bacteria and mitochondria. *Mol Microbiol.* 2000;36:244–246.
11. She Q, Singh RK, Confalonieri F, et al. The complete genome of the crenarchaeon *Sulfolobus solfataricus* P2. *Proc Natl Acad Sci USA.* 2001;98:7835–7840.
12. Jansen R, Embden JD, Gaastra W, Schouls LM. Identification of genes that are associated with DNA repeats in prokaryotes. *Mol Microbiol.* 2002;43:1565–1575.
13. Makarova KS, Aravind L, Grishin NV, Rogozin IB, Koonin EV. A DNA repair system specific for thermophilic Archaea and bacteria predicted by genomic context analysis. *Nucleic Acids Res.* 2002;30:482–496.
14. Haft DH, Selengut J, Mongodin EF, Nelson KE. A guild of 45 CRISPR-associated (Cas) protein families and multiple CRISPR/Cas subtypes exist in prokaryotic genomes. *PLoS Comput Biol.* 2005;1:e60.

15. Makarova KS, Grishin NV, Shabalina SA, Wolf YI, Koonin EV. A putative RNA-interference-based immune system in prokaryotes: computational analysis of the predicted enzymatic machinery, functional analogies with eukaryotic RNAi, and hypothetical mechanisms of action. *Biol Direct.* 2006;1:7.
16. Tang TH, Bachellerie JP, Rozhdestvensky T, Bortolin ML, Huber H, Drungowski M, Elge T, Brosius J, Huttenhofer A. Identification of 86 candidates for small non-messenger RNAs from the archaeon *Archaeoglobus fulgidus. Proc Natl Acad Sci USA.* 2002;99: 7536–7541.
17. Tang TH, Polacek N, Zywicki M, Huber H, Brugger K, Garrett R, Bachellerie JP, Huttenhofer A. Identification of novel non-coding RNAs as potential antisense regulators in the archaeon *Sulfolobus solfataricus. Mol Microbiol.* 2005;55:469–481.
18. Lillestol RK, Shah SA, Brugger K, Redder P, Phan H, Christiansen J, Garrett RA. CRISPR families of the crenarchaeal genus *Sulfolobus*: bidirectional transcription and dynamic properties. *Mol Microbiol.* 2009;72:259–272.
19. Shah SA, Hansen NR, Garrett RA. Distribution of CRISPR spacer matches in viruses and plasmids of crenarchaeal acidothermophiles and implications for their inhibitory mechanism. *Biochem Soc Trans.* 2009;37:23–28.
20. Mojica FJ, Diez-Villasenor C, Garcia-Martinez J, Soria E. Intervening sequences of regularly spaced prokaryotic repeats derive from foreign genetic elements. *J Mol Evol.* 2005;60:174–182.
21. Pourcel C, Salvignol G, Vergnaud G. CRISPR elements in *Yersinia pestis* acquire new repeats by preferential uptake of bacteriophage DNA, and provide additional tools for evolutionary studies. *Microbiology.* 2005;151:653–663.
22. Bolotin A, Quinquis B, Sorokin A, Ehrlich SD. Clustered regularly interspaced short palindrome repeats (CRISPRs) have spacers of extrachromosomal origin. *Microbiology.* 2005;151:2551–2561.
23. Lillestol RK, Redder P, Garrett RA, Brugger K. A putative viral defence mechanism in archaeal cells. *Archaea.* 2006;2:59–72.
24. Barrangou R, Fremaux C, Deveau H, Richards M, Boyaval P, Moineau S, Romero DA, Horvath P. CRISPR provides acquired resistance against viruses in prokaryotes. *Science.* 2007;315:1709–1712.
25. Brouns SJ, Jore MM, Lundgren M, et al. Small CRISPR RNAs guide antiviral defense in prokaryotes. *Science.* 2008;321:960–964.
26. Carte J, Wang R, Li H, Terns RM, Terns MP. Cas6 is an endoribonuclease that generates guide RNAs for invader defense in prokaryotes. *Genes Dev.* 2008;22:3489–3496.
27. Hochstrasser ML, Doudna JA. Cutting it close: CRISPR-associated endoribonuclease structure and function. *Trends Biochem Sci.* 2015;40:58–66.
28. Marraffini LA, Sontheimer EJ. CRISPR interference limits horizontal gene transfer in staphylococci by targeting DNA. *Science.* 2008;322:1843–1845.
29. Hale CR, Zhao P, Olson S, Duff MO, Graveley BR, Wells L, Terns RM, Terns MP. RNA-guided RNA cleavage by a CRISPR RNA-Cas protein complex. *Cell.* 2009;139:945–956.
30. Mojica FJ, Diez-Villasenor C, Garcia-Martinez J, Almendros C. Short motif sequences determine the targets of the prokaryotic CRISPR defence system. *Microbiology.* 2009;155:733–740.
31. Gudbergsdottir S, Deng L, Chen Z, Jensen JV, Jensen LR, She Q, Garrett RA. Dynamic properties of the *Sulfolobus* CRISPR/Cas and CRISPR/Cmr systems when challenged with vector-borne viral and plasmid genes and protospacers. *Mol Microbiol.* 2011;79: 35–49.
32. Westra ER, Semenova E, Datsenko KA, Jackson RN, Wiedenheft B, Severinov K, Brouns SJ. Type I-E CRISPR-cas systems discriminate target from non-target DNA through base pairing-independent PAM recognition. *PLoS Genet.* 2013;9:e1003742.

33. Sashital DG, Wiedenheft B, Doudna JA. Mechanism of foreign DNA selection in a bacterial adaptive immune system. *Mol Cell.* 2012;46:606–615.
34. Semenova E, Jore MM, Datsenko KA, Semenova A, Westra ER, Wanner B, van der Oost J, Brouns SJ, Severinov K. Interference by clustered regularly interspaced short palindromic repeat (CRISPR) RNA is governed by a seed sequence. *Proc Natl Acad Sci USA.* 2011;108:10098–10103.
35. Wiedenheft B, van Duijn E, Bultema JB, et al. RNA-guided complex from a bacterial immune system enhances target recognition through seed sequence interactions. *Proc Natl Acad Sci USA.* 2011;108:10092–10097.
36. Deng L, Garrett RA, Shah SA, Peng X, She Q. A novel interference mechanism by a type IIIB CRISPR-Cmr module in *Sulfolobus. Mol Microbiol.* 2013;87:1088–1099.
37. Marraffini LA, Sontheimer EJ. Self versus non-self discrimination during CRISPR RNA-directed immunity. *Nature.* 2010;463:568–571.
38. Peng W, Feng M, Feng X, Liang YX, She Q. An archaeal CRISPR type III-B system exhibiting distinctive RNA targeting features and mediating dual RNA and DNA interference. *Nucleic Acids Res.* 2015;43:406–417.
39. Garneau JE, Dupuis ME, Villion M, et al. The CRISPR/Cas bacterial immune system cleaves bacteriophage and plasmid DNA. *Nature.* 2010;468:67–71.
40. Deltcheva E, Chylinski K, Sharma CM, Gonzales K, Chao Y, Pirzada ZA, Eckert MR, Vogel J, Charpentier E. CRISPR RNA maturation by *trans*-encoded small RNA and host factor RNase III. *Nature.* 2011;471:602–607.
41. Sapranauskas R, Gasiunas G, Fremaux C, Barrangou R, Horvath P, Siksnys V. The *Streptococcus thermophilus* CRISPR/Cas system provides immunity in *Escherichia coli. Nucleic Acids Res.* 2011;39:9275–9282.
42. Jinek M, Chylinski K, Fonfara I, Hauer M, Doudna JA, Charpentier E. A programmable dual-RNA-guided DNA endonuclease in adaptive bacterial immunity. *Science.* 2012;337:816–821.
43. Gasiunas G, Barrangou R, Horvath P, Siksnys V. Cas9-crRNA ribonucleoprotein complex mediates specific DNA cleavage for adaptive immunity in bacteria. *Proc Natl Acad Sci USA.* 2012;109:E2579–E2586.
44. Nishimasu H, Ran FA, Hsu PD, Konermann S, Shehata SI, Dohmae N, Ishitani R, Zhang F, Nureki O. Crystal structure of Cas9 in complex with guide RNA and target DNA. *Cell.* 2014;156:935–949.
45. Jinek M, Jiang F, Taylor DW, et al. Structures of Cas9 endonucleases reveal RNA-mediated conformational activation. *Science.* 2014;343:1247997.
46. Anders C, Niewoehner O, Duerst A, Jinek M. Structural basis of PAM-dependent target DNA recognition by the Cas9 endonuclease. *Nature.* 2014;513:569–573.
47. Jiang F, Zhou K, Ma L, Gressel S, Doudna JA. Structural biology. A Cas9-guide RNA complex preorganized for target DNA recognition. *Science.* 2015;348:1477–1481.
48. Samai P, Pyenson N, Jiang W, Goldberg GW, Hatoum-Aslan A, Marraffini LA. Co-transcriptional DNA and RNA cleavage during Type III CRISPR-Cas immunity. *Cell.* 2015;161:1164–1174.
49. Tamulaitis G, Venclovas C, Siksnys V. Type III CRISPR-Cas immunity: major differences brushed aside. *Trends Microbiol.* 2017;25:49–61.
50. Taylor DW, Zhu Y, Staals RH, Kornfeld JE, Shinkai A, van der Oost J, Nogales E, Doudna JA. Structural biology. Structures of the CRISPR-Cmr complex reveal mode of RNA target positioning. *Science.* 2015;348:581–585.
51. Staals RH, Agari Y, Maki-Yonekura S, et al. Structure and activity of the RNA-targeting Type III-B CRISPR-Cas complex of *Thermus thermophilus. Mol Cell.* 2013;52:135–145.
52. Tamulaitis G, Kazlauskiene M, Manakova E, Venclovas C, Nwokeoji AO, Dickman MJ, Horvath P, Siksnys V. Programmable RNA shredding by the type III-A CRISPR-Cas system of *Streptococcus thermophilus. Mol Cell.* 2014;56:506–517.

53. Staals RH, Zhu Y, Taylor DW, et al. RNA targeting by the type III-A CRISPR-Cas Csm complex of *Thermus thermophilus*. *Mol Cell.* 2014;56:518–530.
54. Zhu X, Ye K. Cmr4 is the slicer in the RNA-targeting Cmr CRISPR complex. *Nucleic Acids Res.* 2015;43:1257–1267.
55. Elmore JR, Sheppard NF, Ramia N, Deighan T, Li H, Terns RM, Terns MP. Bipartite recognition of target RNAs activates DNA cleavage by the Type III-B CRISPR-Cas system. *Genes Dev.* 2016;30:447–459.
56. Estrella MA, Kuo FT, Bailey S. RNA-activated DNA cleavage by the Type III-B CRISPR-Cas effector complex. *Genes Dev.* 2016;30:460–470.
57. Kazlauskiene M, Tamulaitis G, Kostiuk G, Venclovas C, Siksnys V. Spatiotemporal control of Type III-A CRISPR-Cas immunity: coupling DNA degradation with the target RNA recognition. *Mol Cell.* 2016;62:295–306.
58. Han W, Li Y, Deng L, et al. A type III-B CRISPR-Cas effector complex mediating massive target DNA destruction. *Nucleic Acids Res.* 2016;45:1983–1993.
59. Bondy-Denomy J, Pawluk A, Maxwell KL, Davidson AR. Bacteriophage genes that inactivate the CRISPR/Cas bacterial immune system. *Nature.* 2013;493:429–432.
60. Bondy-Denomy J, Garcia B, Strum S, Du M, Rollins MF, Hidalgo-Reyes Y, Wiedenheft B, Maxwell KL, Davidson AR. Multiple mechanisms for CRISPR-Cas inhibition by anti-CRISPR proteins. *Nature.* 2015;526:136–139.
61. Pawluk A, Bondy-Denomy J, Cheung VH, Maxwell KL, Davidson AR. A new group of phage anti-CRISPR genes inhibits the type I-E CRISPR-Cas system of *Pseudomonas aeruginosa*. *MBio.* 2014;5:e00896.
62. Pawluk A, Staals RH, Taylor C, Watson BN, Saha S, Fineran PC, Maxwell KL, Davidson AR. Inactivation of CRISPR-Cas systems by anti-CRISPR proteins in diverse bacterial species. *Nat Microbiol.* 2016;1:16085.
63. Pawluk A, Amrani N, Zhang Y, et al. Naturally occurring off-switches for CRISPR-Cas9. *Cell.* 2016;167:1829–1838.
64. Rauch BJ, Silvis MR, Hultquist JF, Waters CS, McGregor MJ, Krogan NJ, Bondy-Denomy J. Inhibition of CRISPR-Cas9 with bacteriophage proteins. *Cell.* 2017;168: 150–158.
65. Makarova KS, Wolf YI, Alkhnbashi OS, et al. An updated evolutionary classification of CRISPR-Cas systems. *Nat Rev Microbiol.* 2015;13:722–736.
66. Mohanraju P, Makarova KS, Zetsche B, Zhang F, Koonin EV, van der Oost J. Diverse evolutionary roots and mechanistic variations of the CRISPR-Cas systems. *Science.* 2016;353:aad5147.
67. Zetsche B, Gootenberg JS, Abudayyeh OO, et al. Cpf1 is a single RNA-guided endonuclease of a class 2 CRISPR-Cas system. *Cell.* 2015;163:759–771.
68. Schunder E, Rydzewski K, Grunow R, Heuner K. First indication for a functional CRISPR/Cas system in *Francisella tularensis*. *Int J Med Microbiol.* 2013;303:51–60.
69. Shmakov S, Abudayyeh OO, Makarova KS, et al. Discovery and functional characterization of diverse Class 2 CRISPR-Cas systems. *Mol Cell.* 2015;60:385–397.
70. Abudayyeh OO, Gootenberg JS, Konermann S, et al. C2c2 is a single-component programmable RNA-guided RNA-targeting CRISPR effector. *Science.* 2016;353: aaf5573.
71. Burstein D, Harrington LB, Strutt SC, Probst AJ, Anantharaman K, Thomas BC, Doudna JA, Banfield JF. New CRISPR-Cas systems from uncultivated microbes. *Nature.* 2017;542:237–241.
72. Yang H, Patel DJ. New CRISPR-Cas systems discovered. *Cell Res.* 2017;27:313–314.
73. Cong L, Ran FA, Cox D, et al. Multiplex genome engineering using CRISPR/Cas systems. *Science.* 2013;339:819–823.
74. Mali P, Yang L, Esvelt KM, Aach J, Guell M, DiCarlo JE, Norville JE, Church GM. RNA-guided human genome engineering via Cas9. *Science.* 2013;339:823–826.

75. Gilbert LA, Larson MH, Morsut L, et al. CRISPR-mediated modular RNA-guided regulation of transcription in eukaryotes. *Cell.* 2013;154:442–451.
76. Larson MH, Gilbert LA, Wang X, Lim WA, Weissman JS, Qi LS. CRISPR interference (CRISPRi) for sequence-specific control of gene expression. *Nat Protoc.* 2013;8: 2180–2196.
77. Qi LS, Larson MH, Gilbert LA, Doudna JA, Weissman JS, Arkin AP, Lim WA. Repurposing CRISPR as an RNA-guided platform for sequence-specific control of gene expression. *Cell.* 2013;152:1173–1183.
78. Chen B, Gilbert LA, Cimini BA, et al. Dynamic imaging of genomic loci in living human cells by an optimized CRISPR/Cas system. *Cell.* 2013;155:1479–1491.
79. Hilton IB, D'Ippolito AM, Vockley CM, Thakore PI, Crawford GE, Reddy TE, Gersbach CA. Epigenome editing by a CRISPR-Cas9-based acetyltransferase activates genes from promoters and enhancers. *Nat Biotechnol.* 2015;33:510–517.
80. Kearns NA, Pham H, Tabak B, Genga RM, Silverstein NJ, Garber M, Maehr R. Functional annotation of native enhancers with a Cas9-histone demethylase fusion. *Nat Methods.* 2015;12:401–403.
81. Shalem O, Sanjana NE, Hartenian E, et al. Genome-scale CRISPR-Cas9 knockout screening in human cells. *Science.* 2014;343:84–87.
82. Wang T, Wei JJ, Sabatini DM, Lander ES. Genetic screens in human cells using the CRISPR-Cas9 system. *Science.* 2014;343:80–84.
83. Zhou Y, Zhu S, Cai C, Yuan P, Li C, Huang Y, Wei W. High-throughput screening of a CRISPR/Cas9 library for functional genomics in human cells. *Nature.* 2014;509: 487–491.
84. Koike-Yusa H, Li Y, Tan EP, Velasco-Herrera Mdel C, Yusa K. Genome-wide recessive genetic screening in mammalian cells with a lentiviral CRISPR-guide RNA library. *Nat Biotechnol.* 2014;32:267–273.
85. Gilbert LA, Horlbeck MA, Adamson B, et al. Genome-scale CRISPR-mediated control of gene repression and activation. *Cell.* 2014;159:647–661.
86. Konermann S, Brigham MD, Trevino AE, et al. Genome-scale transcriptional activation by an engineered CRISPR-Cas9 complex. *Nature.* 2015;517:583–588.
87. Jiang W, Marraffini LA. CRISPR-Cas: new tools for genetic manipulations from bacterial immunity systems. *Annu Rev Microbiol.* 2015;69:209–228.
88. Barrangou R, Doudna JA. Applications of CRISPR technologies in research and beyond. *Nat Biotechnol.* 2016;34:933–941.
89. Wang F, Qi LS. Applications of CRISPR genome engineering in cell biology. *Trends Cell Biol.* 2016;26:875–888.
90. Riordan SM, Heruth DP, Zhang LQ, Ye SQ. Application of CRISPR/Cas9 for biomedical discoveries. *Cell Biosci.* 2015;5:33.

CHAPTER TWO

CRISPR/Cas9 Technology: Applications and Human Disease Modeling

Marta Martinez-Lage, Raúl Torres-Ruiz[1], Sandra Rodriguez-Perales[1]
Molecular Cytogenetics and Genome Engineering Group,
Centro Nacional Investigaciones Oncológicas (CNIO), Madrid, Spain
[1] Corresponding author. E-mail address: rtorresr@cnio.es

Contents

Abstract

The CRISPR/Cas9 system development has revolutionized the field of genome engineering through the efficient creation of targeted breaks in the DNA of almost any organism and cell type, opening an avenue for a wide range of applications in biomedical research and medicine. Apart from gene edition through knock-in or knock-out approaches, CRISPR/Cas9 technology has been used for many other purposes, including regulation of endogenous gene expression, epigenome editing, live-cell imaging of chromosomal loci, edition of RNA and high-throughput screening.

Progress in Molecular Biology and Translational Science, Volume 152
ISSN 1877-1173
http://dx.doi.org/10.1016/bs.pmbts.2017.09.002

With all those technological improvements, CRISPR/Cas9 system has broadened the number of alternatives for studying gene function and the generation of more accurate disease models. Although many mechanistic questions remain to be answered and several challenges have yet to be addressed, the use of CRISPR/Cas9-based genome engineering technologies will increase our knowledge of disease processes and their treatment in the near future.

1. GENOME ENGINEERING TOOLS

1.1 Genome Engineering Technologies

Genome editing is an approach in which a target sequence of the genome is directly altered by adding, replacing, or removing DNA bases. Some decades ago, genome-engineering tools were developed in order to create small "cuts" in DNA. This process of cleavage increases the efficiency with which mutations can be induced in target sequences. In the 1980s, Capecchi and Smithies, the pioneers of gene modification, independently developed the technology to modify the genome of mammalian cells by homologous recombination (HR).[1,2] Following on from the discovery of mouse embryonic stem cells (mESCs) by Evans and Kaufman,[3] they developed a process to alter the mouse genome by introducing specific gene modifications in mESCs,[2,4] making this technology a powerful tool for biomedical applications.

For many years, genome engineering was based on HR, but the need for complex targeting strategies and the infrequent occurrence of desired recombination events limited the use of this technology. These limitations drove the development of methods to increase gene-targeting frequency. Advances in the technology were based on the observation that controlled double-strand breaks (DSBs) generated in the DNA increase gene-targeting efficiency and therefore enable the genome to be targeted in a higher frequency.[5–7] Several laboratories demonstrated that a restriction endonuclease (meganuclease) from *Saccharomyces cerevisiae*, I-SceI, generates DSBs and increases targeted HR, with a frequency twofold higher than that of spontaneous HR.[8] Subsequent to this, tools for genome engineering based around programmable DNA-binding regions linked to nonspecific endonucleases were developed.[9] The first targetable nuclease modules were the zinc-finger nucleases (ZFNs),[10] which are based on a bacterial FokI endonuclease domain fused with a zinc-finger domain that recognizes 9 or 18-bp target sequences and cleaves DNA in a sequence-specific manner.[11,12] Few years later this technology was extended by the discovery of transcription activator-like effector nucleases (TALENs).[13,14] Most recently, a programmable-nuclease tool that has truly revolutionized the field is the clustered regularly interspaced short palindromic repeats (CRISPR)/CRISPR-associated (Cas)9 (CRISPR/Cas9) system,[15,16] which was consolidated as a genome-editing tool in 2013 after two groups independently demonstrated efficient edition of mammalian genomes.[17,18]

1.1.1 Homologous Recombination (HR)

HR is a natural mechanism that eukaryotic cells use during meiosis to repair harmful breaks that occur in both strands of DNA, known as DSBs. As its name suggests, HR refers to an exchange of DNA that occurs between large regions of DNA whose sequences are homologous. Early experiments with mammalian somatic cells demonstrated that HR facilitates the integration of a transgene into the genome, and provided evidence that cells possess an efficient enzymatic machinery to mediate HR between exogenous and genomic DNA.[19] Capecchi and coworkers were the first to show that HR of DNA could be induced in vitro in cultured mESCs using homologous sequences of exogenous DNA.[2,4] Together with the findings of Smithies and Evans, this work has been paradigmatic in establishing the field of gene targeting and human disease modeling. Indeed, the discoveries by Mario Capecchi, Martin Evans and Oliver Smithies on the principles of gene modification in mESCs by HR were recognized by the 2007 Nobel Prize in Medicine or Physiology (https://www.nobelprize.org/nobel_prizes/medicine/laureates/2007/).

In HR-based gene targeting approaches, exogenous DNA containing the desired transgene/mutation flanked by segments (arms) of homology with the host endogenous DNA, is integrated into the genome.[2,20] Exogenous DNA is introduced into the cell using standard methods of calcium phosphate-mediated transfection or electroporation, and subsequently recombines into the endogenous DNA, aided by proteins that promote recombination. The length of the homologous DNA arms is a determinant of targeting efficiency, with Bradley and coworkers showing that a length between 1.3 and 6.8 kb is recommended for successful targeting.[21] In addition to homology, the exogenous DNA must contain the sequence of interest along with a positive selection marker, typically an antibiotic resistance cassette, which allows the identification and selection of those cells that have incorporated the exogenous DNA into their genome.

HR-based targeting vectors have been widely used to create knock-out, mutant, and transgenic animals, and have been instrumental in understanding the genetic bases of several diseases.

1.1.2 Double-Strand Breaks (DSBs)

Programmable nucleases generate DSBs, which in turn trigger the host cell machinery to repair the break, predominantly by the nonhomologous end joining (NHEJ) repair mechanism and less frequently by HR. The ability to efficiently introduce a DSB at a target location has made genome-editing technologies a powerful tool to introduce site-specific modifications.[22]

The NHEJ pathway repairs DSBs by adding or deleting bases at the ends of the break without the need for a homologous template. The break is repaired directly by rejoining the DSB ends in an error-prone process, resulting in insertion and deletion (indels) that can cause frameshifts mutations. This mechanism can be exploited to knock-out genes.[23] Alternatively, the homology-directed repair (HDR) pathway can repair the break using homologous sequences in an undamaged sister chromatid, or using an exogenously supplied template with a desired mutation. HDR is the more accurate process, as the DNA template assists in the repair. Nevertheless, HDR with an exogenous template is an inefficient process and only about one in a million cells insert the desired mutation.[24] The exogenous repair template used can be either a double-strand DNA vector or a single-strand DNA oligonucleotide, with the desired mutation flanked by homology arms.

Thus, HDR can be harnessed to introduce a donor template DNA with homologous flanking regions to the DSB region of the target locus to promote reparation with the mutation.[25] This system can be used to generate precise mutations (or knock-ins) by delivery of a DNA donor promoting HDR repair, or can be used for knock-out generation through small nucleotide indels produced by NHEJ repair. Gene-editing methods, such as ZNFs, TALENs, and CRISPR/Cas9 exploit cellular repair mechanisms to allow the manipulation of the genome of many different species.[26]

1.1.3 NHEJ Versus HDR

As mentioned earlier, the production of a DSB triggers the endogenous DNA repair machinery to instigate DNA repair through different pathways: in the case of NHEJ, this can generate small indels; and in the case of HDR, this can lead to error-free DNA repair or the introduction of precise mutations via an exogenous homologous DNA template.[6]

Because DSB healing by NHEJ introduces indels, two-thirds of the generated indels in a coding gene will cause a frameshift that produces a truncated (nonfunctional) protein, in other words, generating a target gene knock-out.[27] Unlike NHEJ, HDR requires the presence of an exogenous DNA repair template to incorporate desired changes in the target gene, which mainly involve the generation of specific mutations.

Although the NHEJ pathway predominates over HR, it has been found that inhibiting key components of the NHEJ pathway, such as DNA ligase IV, increases the proportion of HR-induced products.[28] The HDR pathway can also be promoted by using modified engineered nucleases that cut only one strand, as "nicks" are not a substrate for NHEJ and can stimulates HR; however, this strategy leads to a reduction in efficiency.[29] An alternative approach to suppress NHEJ is the use of small molecules, such as activators of Rad51, a protein involved in HDR. This strategy has been shown to promote HDR,[30] but can also perturb the natural DNA repair mechanisms of the cell, limiting its applicability.

1.1.4 Zinc Finger Nuclease (ZFN)

ZFNs were the first genome-editing tools to use customizable endonucleases. ZFNs are chimeric enzymes comprising a zinc finger (ZF) domain, which binds to DNA, fused to a bacterial cleavage domain from the FokI endonuclease.[11] FokI belongs to the type II family of restriction endonucleases and cleaves a DNA sequence outside of the recognition site. The discovery in 1992 that the cleavage specificity of ZFNs can be modified by linking specific DNA-binding domains, paved the way for their use as gene-editing tools.[31]

For gene-targeting applications, ZFNs are designed in pairs, each recognizing a different DNA strand flanking the genomic target site of interest. ZF domains are protein modules that recognize DNA sequences of 3–4 bp, and can be engineered modularly in tandem to bind a specific DNA sequence of 9–18 bp. At least three consecutive "fingers" are needed to obtain a good binding affinity. Once the ZF domains bind the target DNA target, the fused FokI nuclease domain induces the DSB,[27] which is then repaired by NHEJ or HDR cell repair mechanisms. Subsequent studies showed that the ZF domain can be modified to recognize a sequence of 24 bp or longer that is unique in the genome.[32] Although ZFN technology enables changes to a specific genomic, its use is limited, in part, because targeting new genomic sites requires the creation of new ZF proteins, which is technically challenging.[33] Moreover, ZFNs are restricted in their site-selection as targeting guanine-poor sequences is difficult. The technology is also cost-prohibitive for academic research laboratories.

1.1.5 Transcription Activator-Like Effector Nucleases (TALENs)

TALEN technology was developed about 15 years after the description of ZFNs. Similar to ZFNs, TALENs are chimeric proteins consisting of a modular DNA-binding domain fused to a FokI nuclease that generates a DSB at a specific site.[14,34] TALENs were originally characterized as virulence factors in *Xanthomonas* bacteria.[35] Accordingly, the principal difference between ZNFs and TALENs is that TALENs consist of naturally occurring DNA-binding modules, termed TAL. TALs contain an array of conserved domains of approximately 34 amino acids that differ only at positions 12 and 13, which are the sites of DNA contact and are referred to as repeat variable diresidues (RVDs).[36,37] The RVDs define the DNA-binding specificity of each TALEN module that binds specifically to DNA bases in a 1-RVD to 1-bp ratio. This contrasts with the specificity of ZFNs that recognize DNA triplets.[38,39] Like ZFNs, modular TALEN repeats can be linked together to recognize a specific DNA target sequence. Changing the RVD combination and engineering a tandem array of TAL repeats containing RVDs of interest makes it possible to target different sequences, which is ideal for constructing custom nucleases.[22]

While TALENs represent a significant improvement over ZFNs, their modular nature makes challenging its assembly . A major advantage of using TALENs, however, is that considerably less time is needed to generate a functional nuclease, and there is a significant improvement in specificity and toxicity over ZFNs.[40] By contrast, the large size TALENs presents a challenge for cellular delivery, as they cannot be easily packaged into lentiviruses.[41]

1.1.6 CRISPR/Cas9

It is probably safe to say that the CRISPR/Cas9 system has transformed genome editing. The technology is a modification of the bacterial immune system to create a single guide RNA (sgRNA), composed of a fused CRISP RNA (crRNA) and a *trans*-activating crRNA (tracrRNA),[15] that directs the Cas9 endonuclease (from *Streptococcus pyogenes*) to any site of interest upstream of a protospacer adjacent motif (PAM) sequence. The sgRNA is designed to recognize a target site that matches a 20-bp DNA sequence flanking the three-nucleotide PAM sequence (NGG). Cas9 cleaves the DNA at this site and generates a DSB, which stimulates the activation of NHEJ and HDR (Fig. 1).

The first use of the CRISPR/Cas9 system to induce a DSB in eukaryotic DNA (mouse and human cells) was reported in 2013,[18] and heralded its

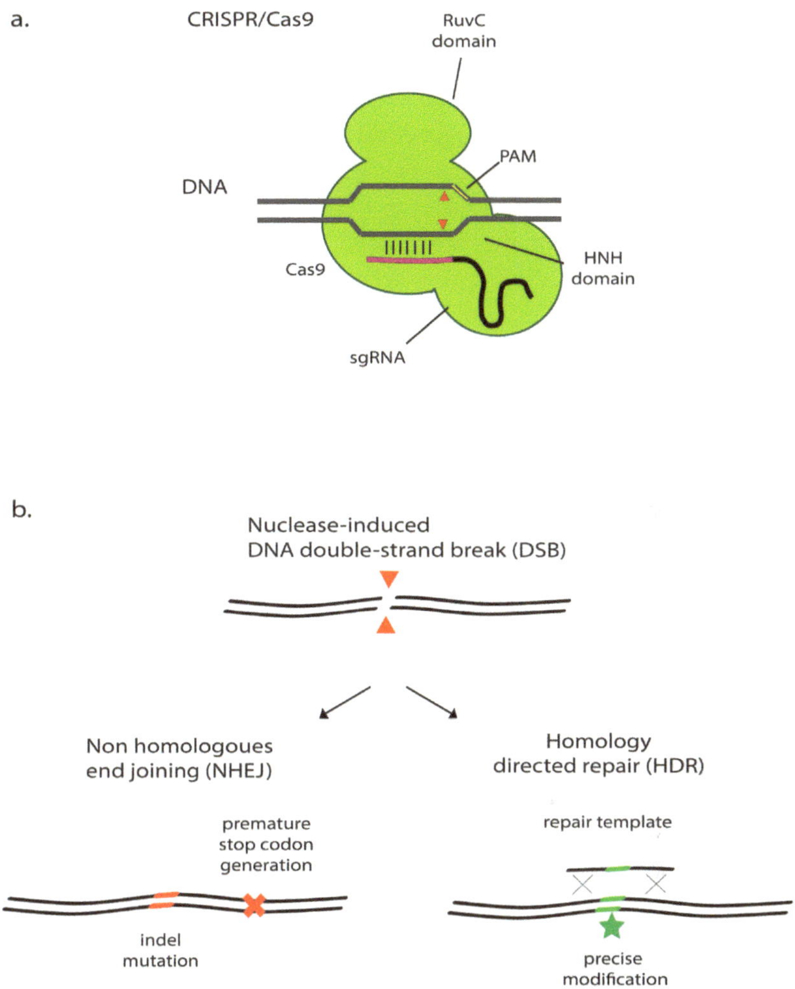

Fig. 1 (A) CRISPR system components. (B) DNA double-strand breaks (DSBs) can follow two different pathways, namely, the error-prone NHEJ repair pathway and the HDR repair pathway, which can be exploited to introduce specific genome alterations. *HDR*, Homology-directed repair; *Indel*, insertion and/or deletion; *NHEJ*, nonhomologous end joining; *PAM*, protospacer adjacent motif; *sgRNA*, single guide RNA.

application for genome editing in eukaryotes. Many reports have subsequently confirmed and extended these findings and have demonstrated the broad utility of the CRISPR/Cas9 system to produce alterations or modifications in model organisms.[42]

CRISPR/Cas9 has enjoyed a rapid development due to its advantages over TALENs and ZFNs, specifically cleavage efficiency, specificity, and simplicity of design. Accordingly, whereas the DNA specificity of ZFNs and TALENs is determined by protein motifs that recognize opposite DNA strands, DNA specificity in CRISPR/Cas9 is determined by Watson-Crick base pairing of the 20-nucleotide long sgRNA. The CRISPR/Cas9 system also has the advantage that only modifying the 20 nucleotides of the sgRNA sequence retargeted it to a different *locus* without the need of protein engineering. On the other hand, TALEN and ZFN platforms require the construction of a new whole protein, which is a far more complex process.[15] A third advantage of the CRISPR/Cas9 system is its precision; while TALENs cleave nonspecifically in the linker between the pair of TALEN-binding sites, CRISPR/Cas9 cleaves 3 bp upstream the PAM sequence.[29]

Despite all of the aforementioned, a limitation in CRISPR/Cas9 technology is that it requires a GG dinucleotide in the PAM sequence, which occurs once in 8 bp in comparison with TALENs that have a target site of 3 bp. Nevertheless, this situation is improving with the use of alternative endonucleases with different PAM specificities,[43] allowing for more flexibility.

The newer CRISPR/Cas9 platforms also overcome ZFN and TALEN limitations by providing a simple way to engineer target DNA breaks, being faster and cheaper. It is thus a promising technology in the genome-engineering field.

1.2 Off Target Effects

A major hurdle common to all the aforementioned technologies is the risk of off-target effects. While the careful design of a genome-editing strategy provides specificity in genetic manipulation, it does not guarantee the absence of off-target cleavage.[44] Off-target mutations can be detected using different approaches that vary in specificity and sensitivity. One technique commonly used to detect off-target mutations is the T7 endonuclease I assay[45]; however, it does detect mutations below 1% and is therefore rather insensitive. To improve upon this, other techniques have been established, including deep-sequencing, web-based prediction tools, and ChIP-seq,[46–48] as well as visualization of multiple repair foci.[9]

Clearly, it is better to minimize the off-target effects at the outset. Accordingly, different strategies have been developed for this, for example, in CRISPR/Cas9 technology, a modified Cas9, Cas9 nickase, has been created that generates a single-strand break in one strand, making it necessary to use two sgRNAs that bind to opposite strands to generate the nick. This strategy improves cleavage specificity while reducing off-target effects.[29] Different methods to detect off-target mutations and ways to reduce them are reviewed in Koo *et al.*[49] and will be explained in more detail in another chapter of this book.

2. CRISPR APPLICATIONS

Soon after the CRISPR/Cas9 system was successfully used to perform targeted genome engineering, many investigators sought to adapt the system to broaden its applications. In addition to enabling efficient genome editing through the creation of targeted DSBs, CRISPR/Cas9 technology has been successfully used for many other purposes (Fig. 2A and B).

2.1 CRISPRi and CRISPRa: Regulation of Transcriptional Levels

One of the first applications of CRISPR/Cas9 outside the field of gene editing was in the regulation of endogenous gene expression. A catalytically inactive version of Cas9 (deadCas9 or dCas9) was generated by double mutations to block enzymatic activities of the RuvC and HNH domains, which cleave the noncomplementary and the complementary strands to the sgRNA, respectively.[50] Inactivation of both nucleases leads to a loss of endonuclease activity, whereas the ability to incorporate sgRNA and bind to targets remains. Consequently, dCas9 can be recruited by sgRNAs to specific target gene promoters and perturb the gene expression without modifying the DNA sequence by blocking transcription initiation and elongation through the dCas9-sgRNA complex. A next step was to obtain a more robust repression of gene expression by fusing dCas9 to transcriptional repressors. Fusing a doxycycline-inducible dCas9 to a Krüppel-associated box (KRAB) repression domain can specifically and reversibly trigger gene silencing. The aforementioned approach, CRISPR interference (CRISPRi), was shown to produce consistent and robust knock-down of gene expression leading to reduced target RNA expression in both human and yeast cells.[51] CRISPRi has also been used as a powerful platform to perform genome-scale screens through the use of several sgRNAs codelivered to silence multiple genes simultaneously. More recently, this approach has been adapted to repress gene expression in a wide variety of organisms and cell types, including bacteria, plants, and human induced pluripotent stem cells (iPSCs).[52–54]

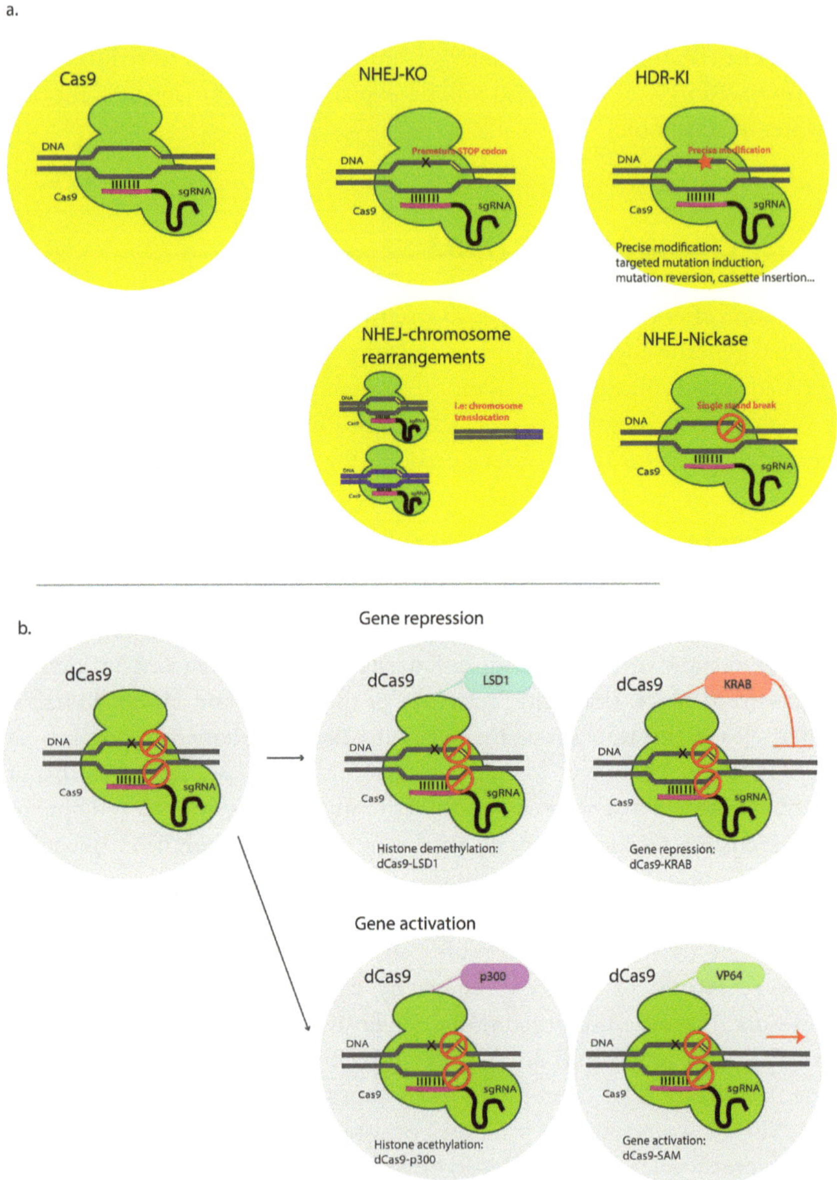

Fig. 2 (A) Main CRISPR/Cas9 system applications that include targeted genome engineering (KO and KI), and the adapted used for many other purposes, such as chromosomal rearrangements. The inactivation of one catalytic domain allowed the generation of a nickase version of Cas9 protein. (B) An alternative use of this system relies on dead Cas9 (dCas9), an inactivated protein that can be fused to functional effectors or domains to mediate fluorescent labeling of DNA loci, transcriptional control, and epigenome modification. *HDR*, Homology-directed repair; *KI*: knock-in; *KO*, knock-out; *NHEJ*, nonhomologous end joining; *sgRNA*, single guide RNA.

As an obvious counter approach, it has been shown that CRISPR-mediated gene activation (CRISPRa) can activate transcription of endogenous genes efficiently through the recruitment of transcription activators by dCas9 fusion proteins.[55–61] The fusion of the VP64 activator domain to dCas9 can activate genes using a single sgRNA[58,61]; however, the use of multiple sgRNAs was found to be necessary to achieve significant activation of endogenous genes.[58,61] The various modifications to this system that have been developed to strengthen activation of transcription include fusion of VP64 to both amino- and carboxy-terminal ends of dCas9,[62] fusion of 10 copies of the minimal VP16 transcriptional activation domain to dCas9, and a synergistic action clustering 3–4 sgRNAs to the promoter.[57,59] CRISPRa also enables robust multiplexed endogenous gene activation when sgRNAs targeting multiple genes are simultaneously introduced into cells.[57] Nevertheless, the complexity of genome-wide activation screens necessitates the use of one efficient sgRNA per gene. The finding that the recruitment of many activators could increase activation efficiency led investigators to study whether dCas9 fused with a carboxy-terminal *SunTag* array, which consists of 10 copies of a small peptide epitope,[63] could increase activation levels. The *SunTag* system relies on the high affinity and specificity of the binding between antibody and antigen to construct a protein-tagging platform to achieve signaling amplification. A cognate single-chain variable fragment (scFV) fused to a superfolder GFP (sfGFP; to improve protein folding) and to VP64 (scFV–sfGFP–VP64) recognized these peptides and recruited multiple copies of VP64 to a single dCas9. Thus, based on the precise guidance of sgRNAs for dCas9 and the distinct binding between antigen and antibody, these transcription regulators could be recruited to the promoter of the target gene, achieving significant activation with a single sgRNA.[63]

2.2 Epigenome Edition

Although all cells have the same DNA sequence, they have the potential to differentiate into many different types that constitute an organism. This remarkable achievement results from the exquisite regulation of the expression of genes and also their access to chromatin. Epigenetic mechanisms, a record of reversible chemical changes in DNA and histones, including histone modifications, DNA methylation, and noncoding RNAs, orchestrate a highly complex system that mediates organization of chromatin structure and epigenetic regulation. The epigenome plays a fundamental role in gene regulation by controlling the expression of regions of the genome, through open or closedconformations.

Epigenome editing is a novel approach whose goal is to rewrite an epigenetic mark at any locus at will, and ultimately modulate the expression of endogenous genes. This type of edition should enable durable gene regulation, with potential applications in basic research and molecular medicine. The CRISPR/Cas9 system can be exploited to perform targeted alteration of chromatin marks and gene expression without modifying the genome sequence. This application has allowed the edition of the epigenome with high spatial and temporal specificity by the fusion of dCas9 to functional domains of DNA methylation or demethylation enzymes, or histone modifiers.[64–67]

To enable eukaryotic epigenome edition, a dCas9-based tool has been created to establish an activating mark. To catalyze the addition of a H3K27 acetylation mark to promoters, proximal enhancers, and distal enhancers, thereby activating gene expression, a catalytic histone acetyltransferase core domain of the human E1A-associated protein p300 was fused to the nuclease-null dCas9.[67] The dSpCas9-p300 module was found to induce a 5- to 270-fold increase in gene expression of downstream genes in the locus control region of the human β-globin gene by directly acetylating distal enhancer regions. However, dSpCas9-VP64 was unable to influence transcription when bound to these same regions, possibly due to the distinct mechanism of activation orchestrated by the epigenetic regulator. To silence transcription through the recruitment of chromatin-modifying complexes, the KRAB domain of Kox1, a naturally-occurring transcriptional repression domain involved in recruiting a heterochromatin-forming domain complex that mediates histone methylation and deacetylation induces gene silencing,[64] was fused to dCas9.[51] Integration of dSpCas9-KRAB into the genome of HEK293 cells resulted in a 15-fold repression of an integrated GFP reporter. These experiments demonstrate that genomic integration of the CRISPR/Cas9 components facilitated long-term silencing, with no detectable GFP expression in 95% of the population after 14 days.

In another approach, the in tandem fusion of four copies of the mSin3 domain (termed SID4X) to the dCas9 scaffold was found to repress the endogenous gene SOX2 in HEK293T cells more than twofold.[68] It has also been shown that histone demethylase LSD1 fused to a small dCas9 ortholog from *N. meningitis* (dNmCas9) can remove activating H3K4 methylation marks, leading to enhancer decommissioning and gene repression.[65]

The next generation of tools to investigate the role of chromatin and DNA modifications should provide a unique opportunity to study fundamental questions in chromatin biology and hold great promise for future personalized medicine approaches.

2.3 Genome Imaging

Imaging modalities offer a direct approach to study the spatial and temporal behavior of the genome in living cells.[69] CRISPR/Cas9 has been repurposed for live-cell labeling of chromosomal loci, facilitating the visualization of chromosomal dynamics, and increasing our understanding of many fundamental intranuclear processes. As an example, inactive dCas9 fused to eGFP has been used to target telomere repetitive elements and various coding genes to enable the live imaging of genomic loci in human cell lines.[70,71] This technique has been used to study the subnuclear localization of the MUC4 loci and the cohesion of replicated MUC4 loci on sister chromatids, and their associated dynamics. In addition, dCas9 fused to eGFP has been used to target pericentric, centromeric, and telomeric repeats in the genome of living mESCs. One of the main caveats of this approach, however, is the sensitivity/intensity of the signal. The development of the *SunTag* system overcomes this issue through the amplification of the signal in the genome by a protein scaffold. This strategy has been used to recruit up to 24 copies of GFP for live imaging.[63]

In 2015, Ma *et al.* engineered a multicolor version of CRISPR/Cas9 using catalytically inactive dCas9 from three different bacterial orthologs[72] to efficiently label various target loci in living human cells. The colored dCas9-sgRNAs were used to determine the intranuclear distance between the loci and the fluorescence spatial resolution in the same chromosome between the two genetic loci. In the same year, a method known as "Cas9-mediated fluorescence in situ hybridization" (CASFISH) was described, which combined the utility of dCas9 with FISH.[73] CASFISH avoids the FISH procedure-related aggressive heat and/or chemical treatments that distort the natural organization of the nucleus, preserving the spatial relationship of the genetic elements. Furthermore, dCas9 from *S. pyogenes* has been successfully used to track endogenous RNA in living cells[47] using sgRNAs targeting mRNAs.[74]

2.4 Large-Scale Functional Genomic Studies Using CRISPR/Cas9

One of the key challenges in basic biology and also in human malignancy studies is clarifying which genes or combination of genes drive natural and malignant processes. The simplicity of sgRNA design and production has opened up the possibility of performing genome-wide targeted mutagenesis, enabling screening of genomic functions through the use of sgRNA libraries. This powerful application of CRISPR technology can be used for genetic screening for cellular phenotypes, and provides considerable improvements over currently employed RNA interference techniques in the specificity of target recognition.[75–78] Combined with the appropriate screening methodologies, this application can be used to simultaneously identify individual or combinations of genes that regulate a variety of phenotypes in eukaryotic organisms and cells.

To generate an sgRNA library, hundreds or thousands of sgRNAs need to be computationally designed and chemically synthesized to target a broad set of genome sequences. These libraries can be used in combination with active Cas9 or inactive dCas9 nucleases, providing a flexible approach to systematically knock-out, activate, or repress hundreds to thousands of genes. One of the key aspects for a successful library construction is the careful design of sgRNAs, including controls, such as several sgRNAs targeting the same gene or promoter region. It is also important to use an efficient and easy-to-control delivery method to ensure the reception of only one sgRNA per cell, usually via lentiviral delivery.[79] Additionally, a carefully chosen screening strategy is essential for successful CRISPR library experiments. The screens can be performed in a pooled manner, coupled to positive or negative selection, or in an arrayed manner. The lentiviral library can be used to transduce cells as a mixed population that is cultured simultaneously. Alternatively, if an automated cell culture system is used, it is possible to study the effect of one specific sgRNA per cell culture. In the pooled approach, a deep sequencing analysis of the sgRNA features in the pooled cells, including genes responsible for cellular phenotypes, such as cell growth and death, can be inferred with bioinformatic tools. The arrayed approach, however, needs only to detect the cell culture with specific phenotypic characteristics to know which genes have been targeted.[80]

CRISPR screens have been used to rapidly identify genes and/or regulatory elements responsible for cell growth, cancer metastasis, or drug resistance.[81] One of the first genome-scale CRISPR knock-out (GeCKO) libraries was reported in 2014,[82] and comprised 64751 sgRNAs targeting 18080 human genes. Using this system, Zhang *et al.* screened candidate genes involved in resistance to vemurafenib, a therapeutic RAF inhibitor for melanoma treatment.[77,82] This approach has also been exploited to screen for genes involved in the DNA mismatch repair pathway and drug resistance to the nucleotide analog 6-thioguanine.[78] Not surprisingly, this is a highly active area; and soon after the first in vitro loss-of-function studies were reported, several in vivo loss-of-function Cas9 screenings investigating a range of phenotypes, including cancer cell drug resistance, and metastasis were published.[83–85] Also, a genome scale loss-of-function library was used to study the genes involved in nonsmall-cell lung cancer metastatic process in a mouse model.[86]

Besides loss-of-function screening, several studies have exploited gain-of-function CRISPR genomic screening, such as transcriptional activation or activation of enhancers.[87–89] In 2014, Gilbert *et al.* constructed genome-scale CRISPRi and CRISPRa libraries to identify essential genes, tumor suppressors, and regulators of differentiation, and to screen for sensitivity to cholera–diphtheria toxin.[87] In 2015, Konermann *et al.* synthesized a library consisting of 70,290 guides targeting all human RefSeq coding isoforms to screen for genes that, upon activation, confer resistance to a BRAF inhibitor.[88] Finally, very recently, Korkmaz *et al.* presented two distinct CRISPR/Cas9 genetic screens to identify and characterize functional enhancers in their native context. Through this analysis, they identified functional enhancer elements that mediate p53 (TP53) and ERα (ESR1) gene regulation.[89]

2.5 Edition of Single-Stranded RNA

A similar molecular "scissors" approach for snipping genes has now been developed to target and cleave RNA. In vitro assays have demonstrated that the Cas9 protein has the ability to edit single-stranded RNA. In 2016, Nelles *et al.* demonstrated that a *S. pyogenes* inactive Cas9 can bind RNA, allowing endogenous RNA tracking in living cells in a programmable manner without genetically encoded tags.[74] Also in 2016, Zhang and coworkers characterized the class 2 type VI CRISPR-Cas effector C2c2 from the bacterium *Leptotrichia shahii* and demonstrated its RNA-guided ribonuclease function. C2c2 can be programmed to cleave single-stranded RNA targets carrying complementary protospacers.[90] This CRISPR application can be used to develop new RNA-targeting tools.

With an ever-increasing array of possible uses, CRISPR/Cas9-based technologies have captured the imagination of biologists, and rightly so.

Aside from the earlier mentioned applications that have emerged in the last 4 years, the CRISPR/Cas9 system has been used successfully to model many human diseases that until recently were impossible to mimic with the preexisting technology. In the next section we review the most important advances made in this regard.

3. CRISPR APPLICATIONS IN BIOMEDICAL MODELING

One of the major hurdles in the generation of faithful disease models is not only the recreation of patient-specific mutations or alterations, but also the use of proper controls.[91] With regard to this, the CRISPR/Cas9 system has facilitated the reconstruction of both phenotypes (disease-related models or gene correction alternatives). In a very short space of time, the CRISPR/Cas9 system has been used to generate many disease-based models for many important human pathologies, including neurological diseases[92] cancer,[93] and cardiovascular pathologies,[94] as well as other Mendelian or complex genetic human diseases, which has allowed the investigation of the molecular mechanisms underlying pathogenesis. These new models are also excellent platforms for drug screening, high-throughput studies, and gene therapy purposes.

Here we summarize, in a nonextensive manner, what we believe are the most important reports that been published in the last year in relation to disease modeling.

3.1 Neurological Disease Models

The use of the CRISPR/Cas9 system as an error-prone mechanism (based on its ability to generate DSBs and their repair by NHEJ) has enabled the generation of human disease models in a fast, efficient, and cost-effective manner. This has permitted the introduction of mutations in several neuronal loci in zebrafish,[95] and in mice and rats.[96]

A second application has focused on the use of the CRISPR/Cas9 system to engineer iPSCs and generate animal models, specifically by exploiting the HDR repair pathway. Two good examples of this application are: (i) the use of an iPSC-based model to explore the underlying mechanism of epilepsy caused by SCN1A loss-of-function mutations.[97] To distinguish GABAergic subtype neurons derived from edited iPSCs and the iPSCs of patients and controls, the authors used a knock-in strategy to fluorescently label differentiated neurons. This approach showed that Nav1.1 was expressed primarily on GABAergic neurons but rarely on glutamatergic neurons. (ii) The use of patient-derived iPSCs to establish cell-autonomous disease models of

Parkinson disease (PD),[98] Alzheimer disease,[99] and Huntington disease (HD),[100] among others.

A third application has used CRISP/Cas9 to induce exon deletion and restore gene functionality. Tabebordbar *et al.* developed a dual CRISPR approach to directly edit muscle stem cells, and also an in vivo approach to correct mouse muscles by disrupting the exon-coding region that harbors pathogenic mutations in the *Dmd* (Duchenne muscular dystrophy) gene. The in vivo delivery of Cas9 by adeno-associated virus (AAV) coupled with two different sgRNAs targeting both ends of exon 23 in the mutated *Dmd* gene promoted the deletion of the exon, thus producing a truncated, but functional protein. This treatment partially resolved muscle functional deficiencies.

3.2 Cancer Models

Cancer is one of the most common causes of death in first world countries. There are numerous ways in which this devastating pathology can be studied, for example, by using patient-derived cells, animal models (either by generating the pathology or by using them as a xenograft model), or cooperative studies with patients. Some of the chief problems with these methodologies is the use of late-stage samples (in the case of patient-derived or patient studies), or the difficult recapitulation of human pathology or primary oncogenic events.[101] With regard to this, numerous studies have been carried out with the aim of generating accurate and specific in vivo and in vitro cancer models.[93] In a study by Maresch *et al.*,[102] the authors reported the generation of a pancreatic cancer model by simultaneous edition of multiple gene network sets using transfection-based multiplex delivery of CRISPR/Cas9 components to the pancreas of adult mice. In the same study, the authors modeled complex chromosomal rearrangements, a hallmark of pancreatic cancer. In 2014, Sanchez-Rivera *et al.*[103] described an approach for functional characterization of candidate genes in mouse models of cancer. As proof-of-concept, they used a *Kras* (G12D)-driven lung cancer model, and used CRISPR/Cas9 to edit the genome of tumor suppressor genes with known loss-of-function alterations in human lung cancer, resulting in the generation of lung adenocarcinomas in mice. Besides these in vivo models, there are several examples of in vitro models capable of recapitulating the first stages of specific human cancers, which until now have been elusive to obtain, such as brain tumors,[104] Ewing sarcoma,[105] and colorectal cancer,[106] among others.

3.3 Cardiovascular Disease Models

CRISPR/Cas9 has been used to increase our knowledge of electrophysiological disorders,[107,108] lipid metabolism,[109] and cardiomyopathies.[110] In a recent report, Carroll *et al.*[111] generated transgenic cardiac-specific Cas9 mice by injecting a Cas9 expression plasmid regulated by the *Myh6* promoter into mouse zygotes. The transgenic mice robustly expressed high levels of Cas9 exclusively in heart cardiomyocytes. As a proof-of-concept experiment, the authors used AAV to deliver sgRNAs against *Myh6*, and demonstrated cardiac-specific genome editing at this locus. In 2014, Ding *et al.*[112] introduced targeted loss-of-function mutations into the endogenous *Pcsk9* gene using adenovirus-delivered Cas9 and sgRNAs targeting *Pcsk9* in mouse liver. The authors reported a high rate of mutagenesis (around 50%), and an analysis of related pathophysiological effects revealed a reduction in blood cholesterol levels.

3.4 Infectious Disease Models

CRISPR/Cas9 system can be easily harnessed to treat viral infections due to the specific genomic nature of the foreign agents. One study[113] has taken advantage of these characteristics by expressing CRISPR/Cas9 in transduced cells via lentiviral genome integration. CRISPR has also been used to specifically target HIV in latently-infected T cell lines and in the cellular reservoirs of HIV, providing long-term resistance to HIV-1.[114] It has also been demonstrated that the system can target and cleave conserved regions in the chronic hepatitis B virus after infection, resulting in strong suppression of viral gene expression and replication[115]; likewise, CRISPR/Cas9 has been used to recognize specifically HSV-1 viral genomes to inhibit viral replication.[116]

3.5 Immunodeficiency Models

In 2014, Huang and coworkers[117] combined the multiplex characteristics of CRISPR and embryo microinjection with multiple sgRNAs and Cas9 mRNA. Using this approach, they were able to target *B2m*, *Il2rg*, *Prf1*, *Prkdc*, and *Rag1*, generating many different immunodeficient mice. In a more recent study, Hoodless *et al.*[118] were able to demonstrate that CDK9 is involved in the resolution of neutrophil-dependent inflammation using pharmacological and CRISPR approaches.

4. CONCLUDING REMARKS

CRISPR/Cas9 has revolutionized biomedical science not only by enabling genome modifications at single-nucleotide resolution in almost any cell type and organism, but also by the fact that it is a fast, easy, and efficient tool that is able to generate what was previously thought to be impossible. Furthermore, an experiment based on the CRISPR/Cas9 system can be performed using only a plasmid that expresses the nuclease and an sgRNA targeting a specific locus. CRISPR technology enables researchers to engineer more animals in more complex ways and in a wider range of species. The widespread use of complementary applications based on CRISPR have made it a multifunctional platform whose applicability goes beyond conventional single-gene edition to encompass multiplexed edition, sequence-specific regulation of gene expression, and genome-wide screens of other platforms. These new methodological developments have considerably increased the technological alternatives for studying gene function and for modeling in several organisms and diseases. Moreover, the combination of CRISPR-based genome engineering and genome-wide association studies could play a key role in the development of personalized therapy.

In 2015, the first study on the use of CRISPR/Cas9 in human zygotes was published. Although the zygotes were unable to develop into viable embryos (they were triploid), the study raised ethical alarms and led to a worldwide moratorium on genomic engineering of human germlines by both biologists and ethicists.[119] The rapid advances in CRISPR/Cas9 systems toward application for treatment of human diseases have given rise to the discussion of the ethical implications of this technology.[120] Before human germlines can be genetically engineered, appropriate control models are warranted for testing efficacy and safety, to minimize collateral effects. Such models would include more accurate and sensitive tools to assess off-target events and mosaicism.[121]

CRISPR/Cas9 is a groundbreaking technology. It heralds an era of changes, with potential application in therapeutics, disease modeling, and genetic studies. Efforts are already under way to develop CRISPR/Cas9-based treatments for cancers of all levels of genetic complexity. The challenge facing researchers today is to develop innovative technologies and improve the safety and efficacy of the new tools. CRISPR models are revealing complex relationships with wide-ranging implications for human diseases. The combination of such approaches with next-generation sequencing data provides an unprecedented opportunity to create powerful and more informative cellular and animal genetic models that enhance our understanding of the pathological processes. We believe that many of the applications of CRISPR-based genome engineering technology will help to decipher both intrinsic and microenvironmental cellular elements that play a role in many diseases, their progression, and may even enable a cure for some of them.

REFERENCES

1. Smithies O, Gregg RG, Boggs SS, Koralewski MA, Kucherlapati RS. Insertion of DNA sequences into the human chromosomal beta-globin locus by homologous recombination. *Nature.* 1985;317(6034):230–234.
2. Thomas KR, Capecchi MR. Introduction of homologous DNA sequences into mammalian cells induces mutations in the cognate gene. *Nature.* 1986;324(6092):34–38.
3. Evans MJ, Kaufman MH. Establishment in culture of pluripotential cells from mouse embryos. *Nature.* 1981;292(5819):154–156.
4. Thomas KR, Capecchi MR. Site-directed mutagenesis by gene targeting in mouse embryo-derived stem cells. *Cell.* 1987;51(3):503–512.
5. Rouet P, Smih F, Jasin M. Introduction of double-strand breaks into the genome of mouse cells by expression of a rare-cutting endonuclease. *Mol Cell Biol.* 1994;14 (12):8096–8106.
6. Jasin M. Genetic manipulation of genomes with rare-cutting endonucleases. *Trends Genet.* 1996;12(6):224–228.
7. Rudin N, Sugarman E, Haber JE. Genetic and physical analysis of double-strand break repair and recombination in *Saccharomyces cerevisiae*. *Genetics.* 1989;122(3):519–534.
8. Choulika A, Perrin A, Dujon B, Nicolas JF. Induction of homologous recombination in mammalian chromosomes by using the I-SceI system of *Saccharomyces cerevisiae*. *Mol Cell Biol.* 1995;15(4):1968–1973.
9. Chandrasegaran S, Carroll D. Origins of programmable nucleases for genome engineering. *J Mol Biol.* 2016;428(5 Pt B):963–989.
10. Bibikova M, Beumer K, Trautman JK, Carroll D. Enhancing gene targeting with designed zinc finger nucleases. *Science.* 2003;300(5620):764–774.
11. Kim YG, Cha J, Chandrasegaran S. Hybrid restriction enzymes: zinc finger fusions to Fok I cleavage domain. *Proc Natl Acad Sci USA.* 1996;93(3):1156–1160.
12. Beerli RR, Segal DJ, Dreier B, Barbas CF. Toward controlling gene expression at will: specific regulation of the erbB-2/HER-2 promoter by using polydactyl zinc finger proteins constructed from modular building blocks. *Proc Natl Acad Sci USA.* 1998;95 (25):14628–14633.
13. Moscou MJ, Bogdanove AJ. A simple cipher governs DNA recognition by TAL effectors. *Science.* 2009;326(5959):1501–1511.
14. Christian M, Cermak T, Doyle EL, Schmidt C, Zhang F, Hummel A, et al. Targeting DNA double-strand breaks with TAL effector nucleases. *Genetics.* 2010;186(2): 757–761.
15. Jinek M, Chylinski K, Fonfara I, Hauer M, Doudna JA, Charpentier E. A programmable dual-RNA-guided DNA endonuclease in adaptive bacterial immunity. *Science.* 2012;337(6096):816–821.
16. Doudna JA, Charpentier E. Genome editing. The new frontier of genome engineering with CRISPR-Cas9. *Science.* 2014;346(6213):1258096–1258106.

17. Mali P, Yang L, Esvelt KM, Aach J, Guell M, DiCarlo JE, et al. RNA-guided human genome engineering via Cas9. *Science.* 2013;339(6121):823–826.
18. Cong L, Ran FA, Cox D, Lin S, Barretto R, Habib N, et al. Multiplex genome engineering using CRISPR/Cas systems. *Science.* 2013;339(6121):819–823.
19. Folger KR, Wong EA, Wahl G, Capecchi MR. Patterns of integration of DNA microinjected into cultured mammalian cells: evidence for homologous recombination between injected plasmid DNA molecules. *Mol Cell Biol.* 1982;2(11): 1372–1387.
20. Thomas KR, Folger KR, Capecchi MR. High frequency targeting of genes to specific sites in the mammalian genome. *Cell.* 1986;44(3):419–428.
21. Hasty P, Rivera-Pérez J, Bradley A. The length of homology required for gene targeting in embryonic stem cells. *Mol Cell Biol.* 1991;11(11):5586–5591.
22. Gaj T, Sirk SJ, Shui S-L, Liu J. Genome-editing technologies: principles and applications. *Cold Spring Harb Perspect Biol.* 2016;8(12):a023754.
23. Honma M, Sakuraba M, Koizumi T, Takashima Y, Sakamoto H, Hayashi M. Non-homologous end-joining for repairing I-SceI-induced DNA double strand breaks in human cells. *DNA Repair.* 2007;6(6):781–788.
24. Kim H, Kim J-S. A guide to genome engineering with programmable nucleases. *Nat Rev Genet.* 2014;15(5):321–334.
25. Ran FA, Hsu PD, Wright J, Agarwala V, Scott DA, Zhang F. Genome engineering using the CRISPR-Cas9 system. *Nat Protoc.* 2013;8(11):2281–2308.
26. Carroll D. Genome engineering with targetable nucleases. *Annu Rev Biochem.* 2014;83 (1):409–439.
27. Urnov FD, Rebar EJ, Holmes MC, Zhang HS, Gregory PD. Genome editing with engineered zinc finger nucleases. *Nat Rev Genet.* 2010;11(9):636–646.
28. Bozas A, Beumer KJ, Trautman JK, Carroll D. Genetic analysis of zinc-finger nuclease-induced gene targeting in *Drosophila. Genetics.* 2009;182(3):641–651.
29. Ran FA, Hsu PD, Lin C-Y, Gootenberg JS, Konermann S, Trevino AE, et al. Double nicking by RNA-guided CRISPR Cas9 for enhanced genome editing specificity. *Cell.* 2013;154(6):1380–1389.
30. Song J, Yang D, Xu J, Zhu T, Chen YE, Zhang J. RS-1 enhances CRISPR/Cas9- and TALEN-mediated knock-in efficiency. *Nat Commun.* 2016;7:10548.
31. Li L, Wu LP, Chandrasegaran S. Functional domains in Fok I restriction endonuclease. *PNAS.* 1992;89(10):4275–4279.
32. Kim HJ, Lee HJ, Kim H, Cho SW, Kim J-S. Targeted genome editing in human cells with zinc finger nucleases constructed via modular assembly. *Genome Res.* 2009;19 (7):1279–1288.
33. Ramirez CL, Foley JE, Wright DA, Müller-Lerch F, Rahman SH, Cornu TI, et al. Unexpected failure rates for modular assembly of engineered zinc fingers. *Nat Methods.* 2008;5(5):374–375.
34. Miller JC, Tan S, Qiao G, Barlow KA, Wang J, Xia DF, et al. A TALE nuclease architecture for efficient genome editing. *Nat Biotechnol.* 2011;29(2):143–148.
35. Boch J, Scholze H, Schornack S, Landgraf A, Hahn S, Kay S, et al. Breaking the code of DNA binding specificity of TAL-type III effectors. *Science.* 2009;326(5959): 1509–1512.
36. Richter A, Streubel J, Boch J. TAL effector DNA-binding principles and specificity. *Methods Mol Biol.* 2016;1338:9–25[Chapter 2].
37. Deng D, Yan C, Pan X, Mahfouz M, Wang J, Zhu J-K, et al. Structural basis for sequence-specific recognition of DNA by TAL effectors. *Science.* 2012;335(6069): 720–723.
38. Lin J, Musunuru K. Genome engineering tools for building cellular models of disease. *FEBS J.* 2016;283(17):3222–3231.

39. Mak AN., Bradley P, Bogdanove AJ, Stoddard BL. TAL effectors: function, structure, engineering and applications. *Curr Opin Struct Biol.* 2013;23(1):93–99.
40. Mussolino C, Alzubi J, Fine EJ, Morbitzer R, Cradick TJ, Lahaye T, et al. TALENs facilitate targeted genome editing in human cells with high specificity and low cytotoxicity. *Nucl Acids Res.* 2014;42(10):6762–6773.
41. Holkers M, Maggio I, Liu J, Janssen JM, Miselli F, Mussolino C, et al. Differential integrity of TALE nuclease genes following adenoviral and lentiviral vector gene transfer into human cells. *Nucl Acids Res.* 2013;41(5):e63–e73.
42. Sander JD, Joung JK. CRISPR-Cas systems for editing, regulating and targeting genomes. *Nat Biotechnol.* 2014;32(4):347–355.
43. Kleinstiver BP, Prew MS, Tsai SQ, Topkar VV, Nguyen NT, Zheng Z, et al. Engineered CRISPR-Cas9 nucleases with altered PAM specificities. *Nature.* 2015;523(7561):481–485.
44. Lee CM, Cradick TJ, Fine EJ, Bao G. Nuclease target site selection for maximizing on-target activity and minimizing off-target effects in genome editing. *Mol Ther.* 2016;24(3):475–487.
45. Mashal RD, Koontz J, Sklar J. Detection of mutations by cleavage of DNA heteroduplexes with bacteriophage resolvases. *Nat Genet.* 1995;9(2):177–183.
46. Kim D, Bae S, Park J, Kim E, Kim S, Yu HR, et al. Digenome-seq: genome-wide profiling of CRISPR-Cas9 off-target effects in human cells. *Nat Methods.* 2015;12(3). [237–243–1pfollowing243].
47. Kuscu C, Arslan S, Singh R, Thorpe J, Adli M. Genome-wide analysis reveals characteristics of off-target sites bound by the Cas9 endonuclease. *Nat Biotechnol.* 2014;32(7):677–683.
48. Singh R, Kuscu C, Quinlan A, Qi Y, Adli M. Cas9-chromatin binding information enables more accurate CRISPR off-target prediction. *Nucl Acids Res.* 2015;43(18):e118.
49. Koo T, Lee J, Kim J-S. Measuring and reducing off-target activities of programmable nucleases including CRISPR-Cas9. *Mol Cells.* 2015;38(6):475–481.
50. Wu X, Scott DA, Kriz AJ, Chiu AC, Hsu PD, Dadon DB, et al. Genome-wide binding of the CRISPR endonuclease Cas9 in mammalian cells. *Nat Biotechnol.* 2014;32(7):670–676.
51. Gilbert LA, Larson MH, Morsut L, Liu Z, Brar GA, Torres SE, et al. CRISPR-mediated modular RNA-guided regulation of transcription in eukaryotes. *Cell.* 2013;154(2):442–451.
52. Mandegar MA, Huebsch N, Frolov EB, Shin E, Truong A, Olvera MP, et al. CRISPR interference efficiently induces specific and reversible gene silencing in human iPSCs. *Cell Stem Cell.* 2016;18(4):541–553.
53. Vazquez-Vilar M, Bernabé-Orts JM, Fernandez-Del-Carmen A, Ziarsolo P, Blanca J, Granell A, et al. A modular toolbox for gRNA-Cas9 genome engineering in plants based on the GoldenBraid standard. *Plant Methods.* 2016;12(1):10.
54. Yao L, Cengic I, Anfelt J, Hudson EP. Multiple gene repression in *Cyanobacteria* using CRISPRi. *ACS Synth Biol.* 2016;5(3):207–212.
55. Chavez A, Scheiman J, Vora S, Pruitt BW, Tuttle M, P R Iyer E, et al. Highly efficient Cas9-mediated transcriptional programming. *Nat Methods.* 2015;12(4):326–328.
56. Kearns NA, Genga RMJ, Enuameh MS, Garber M, Wolfe SA, Maehr R. Cas9 effector-mediated regulation of transcription and differentiation in human pluripotent stem cells. *Development.* 2014;141(1):219–223.
57. Cheng AW, Wang H, Yang H, Shi L, Katz Y, Theunissen TW, et al. Multiplexed activation of endogenous genes by CRISPR-on, an RNA-guided transcriptional activator system. *Cell Res.* 2013;23(10):1163–1171.

58. Maeder ML, Linder SJ, Cascio VM, Fu Y, Ho QH, Joung JK. CRISPR RNA-guided activation of endogenous human genes. *Nat Methods.* 2013;10(10):977–979.
59. Chakraborty S, Ji H, Kabadi AM, Gersbach CA, Christoforou N, Leong KW. A CRISPR/Cas9-based system for reprogramming cell lineage specification. *Stem Cell Rep.* 2014;3(6):940–947.
60. Mali P, Aach J, Stranges PB, Esvelt KM, Moosburner M, Kosuri S, et al. CAS9 transcriptional activators for target specificity screening and paired nickases for cooperative genome engineering. *Nat Biotechnol.* 2013;31(9):833–838.
61. Perez-Pinera P, Kocak DD, Vockley CM, Adler AF, Kabadi AM, Polstein LR, et al. RNA-guided gene activation by CRISPR-Cas9-based transcription factors. *Nat Methods.* 2013;10(10):973–976.
62. Gao X, Tsang JCH, Gaba F, Wu D, Lu L, Liu P. Comparison of TALE designer transcription factors and the CRISPR/dCas9 in regulation of gene expression by targeting enhancers. *Nucl Acids Res.* 2014;42(20):e155–e165.
63. Tanenbaum ME, Gilbert LA, Qi LS, Weissman JS, Vale RD. A protein-tagging system for signal amplification in gene expression and fluorescence imaging. *Cell.* 2014;159 (3):635–646.
64. Thakore PI, D'Ippolito AM, Song L, Safi A, Shivakumar NK, Kabadi AM, et al. Highly specific epigenome editing by CRISPR-Cas9 repressors for silencing of distal regulatory elements. *Nat Methods.* 2015;12(12):1143–1149.
65. Kearns NA, Pham H, Tabak B, Genga RM, Silverstein NJ, Garber M, et al. Functional annotation of native enhancers with a Cas9-histone demethylase fusion. *Nat Methods.* 2015;12(5):401–403.
66. Thakore PI, Black JB, Hilton IB, Gersbach CA. Editing the epigenome: technologies for programmable transcription and epigenetic modulation. *Nat Methods.* 2016;13 (2):127–137.
67. Hilton IB, D'Ippolito AM, Vockley CM, Thakore PI, Crawford GE, Reddy TE, et al. Epigenome editing by a CRISPR-Cas9-based acetyltransferase activates genes from promoters and enhancers. *Nat Biotechnol.* 2015;33(5):510–517.
68. Konermann S, Brigham MD, Trevino A, Hsu PD, Heidenreich M, Cong L, et al. Optical control of mammalian endogenous transcription and epigenetic states. *Nature.* 2013;500(7463):472–476.
69. Smith C, Gore A, Yan W, Abalde-Atristain L, Li Z, He C, et al. Whole-genome sequencing analysis reveals high specificity of CRISPR/Cas9 and TALEN-based genome editing in human iPSCs. *Cell Stem Cell.* 2014;15(1):12–13.
70. Chen B, Gilbert LA, Cimini BA, Schnitzbauer J, Zhang W, Li G-W, et al. Dynamic imaging of genomic loci in living human cells by an optimized CRISPR/Cas system. *Cell.* 2013;155(7):1479–1491.
71. Chen B, Hu J, Almeida R, Liu H, Balakrishnan S, Covill-Cooke C, et al. Expanding the CRISPR imaging toolset with *Staphylococcus aureus* Cas9 for simultaneous imaging of multiple genomic loci. *Nucl Acids Res.* 2016;44:e75.
72. Ma H, Naseri A, Reyes-Gutierrez P, Wolfe SA, Zhang S, Pederson T. Multicolor CRISPR labeling of chromosomal loci in human cells. *Proc Natl Acad Sci USA.* 2015;112(10):3002–3007.
73. Deng W, Shi X, Tjian R, Lionnet T, Singer RH. CASFISH: CRISPR/Cas9-mediated in situ labeling of genomic loci in fixed cells. *Proc Natl Acad Sci USA.* 2015;112 (38):11870–11875.
74. Nelles DA, Fang MY, O'Connell MR, Xu JL, Markmiller SJ, Doudna JA, et al. Programmable RNA tracking in live cells with CRISPR/Cas9. *Cell.* 2016;165(2): 488–496.
75. Ma Y, Creanga A, Lum L, Beachy PA. Prevalence of off-target effects in *Drosophila* RNA interference screens. *Nature.* 2006;443(7109):359–363.

76. Koike-Yusa H, Li Y, Tan E-P, Velasco-Herrera MDC, Yusa K. Genome-wide recessive genetic screening in mammalian cells with a lentiviral CRISPR-guide RNA library. *Nat Biotechnol.* 2014;32(3):267–273.
77. Shalem O, Sanjana NE, Hartenian E, Shi X, Scott DA, Mikkelson T, et al. Genome-scale CRISPR-Cas9 knockout screening in human cells. *Science.* 2014;343(6166): 84–87.
78. Wang T, Wei JJ, Sabatini DM, Lander ES. Genetic screens in human cells using the CRISPR-Cas9 system. *Science.* 2014;343(6166):80–84.
79. Bassett AR, Kong L, Liu J-L. A genome-wide CRISPR library for high-throughput genetic screening in *Drosophila* cells. *J Genet Genomics.* 2015;42(6):301–309.
80. Agrotis A, Ketteler R. A new age in functional genomics using CRISPR/Cas9 in arrayed library screening. *Front Genet.* 2015;6(e51942):300.
81. Kotterman MA, Schaffer DV. Engineering adeno-associated viruses for clinical gene therapy. *Nat Rev Genet.* 2014;15(7):445–451.
82. Niu Y, Shen B, Cui Y, Chen Y, Wang J, Wang L, et al. Generation of gene-modified cynomolgus monkey via Cas9/RNA-mediated gene targeting in one-cell embryos. *Cell.* 2014;156(4):836–843.
83. Barretina J, Caponigro G, Stransky N, Venkatesan K, Margolin AA, Kim S, et al. The Cancer Cell Line Encyclopedia enables predictive modelling of anticancer drug sensitivity. *Nature.* 2012;483(7391):603–607.
84. Lin B-R, Natarajan V. Negative regulation of human U6 snRNA promoter by p38 kinase through Oct-1. *Gene.* 2012;497(2):200–207.
85. Sanjana NE, Shalem O, Zhang F. Improved vectors and genome-wide libraries for CRISPR screening. *Nat Methods.* 2014;11(8):783–784.
86. Chen S, Sanjana NE, Zheng K, Shalem O, Lee K, Shi X, et al. Genome-wide CRISPR screen in a mouse model of tumor growth and metastasis. *Cell.* 2015;160(6):1246–1260.
87. Gilbert LA, Horlbeck MA, Adamson B, Villalta JE, Chen Y, Whitehead EH, et al. Genome-scale CRISPR-mediated control of gene repression and activation. *Cell.* 2014;159(3):647–661.
88. Konermann S, Brigham MD, Trevino AE, Joung J, Abudayyeh OO, Barcena C, et al. Genome-scale transcriptional activation by an engineered CRISPR-Cas9 complex. *Nature.* 2015;517(7536):583–588.
89. Korkmaz G, Lopes R, Ugalde AP, Nevedomskaya E, Han R, Myacheva K, et al. Functional genetic screens for enhancer elements in the human genome using CRISPR-Cas9. *Nat Biotechnol.* 2016;34(2):192–198.
90. Abudayyeh OO, Gootenberg JS, Konermann S, Joung J, Slaymaker IM, Cox DBT, et al. C2c2 is a single-component programmable RNA-guided RNA-targeting CRISPR effector. *Science.* 2016;353(6299):aaf5573.
91. Soldner F, Laganière J, Cheng AW, Hockemeyer D, Gao Q, Alagappan R, et al. Generation of isogenic pluripotent stem cells differing exclusively at two early onset Parkinson point mutations. *Cell.* 2011;146(2):318–331.
92. Heidenreich M, Zhang F. Applications of CRISPR-Cas systems in neuroscience. *Nat Rev Neurosci.* 2016;17(1):36–44.
93. Torres-Ruiz R, Rodriguez-Perales S. CRISPR-Cas9: a revolutionary tool for cancer modelling. *Int J Mol Sci.* 2015;16(9):22151–22168.
94. Seeger T, Porteus M, Wu JC. Genome editing in cardiovascular biology. *Circ Res.* 2017;120(5):778–780.
95. Hwang WY, Fu Y, Reyon D, Maeder ML, Tsai SQ, Sander JD, et al. Efficient genome editing in zebrafish using a CRISPR-Cas system. *Nat Biotechnol.* 2013;31(3):227–229.
96. Li D, Qiu Z, Shao Y, Chen Y, Guan Y, Liu M, et al. Heritable gene targeting in the mouse and rat using a CRISPR-Cas system. *Nat Biotechnol.* 2013;31(8):681–683.

97. Liu J, Gao C, Chen W, Ma W, Li X, Shi Y, et al. CRISPR/Cas9 facilitates investigation of neural circuit disease using human iPSCs: mechanism of epilepsy caused by an SCN1A loss-of-function mutation. *Transl Psychiatry.* 2016;6(1):e703.
98. Zhou X, Xin J, Fan N, Zou Q, Huang J, Ouyang Z, et al. Generation of CRISPR/Cas9-mediated gene-targeted pigs via somatic cell nuclear transfer. *Cell Mol Life Sci.* 2015;72(6):1175–1184.
99. Paquet D, Kwart D, Chen A, Sproul A, Jacob S, Teo S, et al. Efficient introduction of specific homozygous and heterozygous mutations using CRISPR/Cas9. *Nature.* 2016;533(7601):125–129.
100. An MC, O'Brien RN, Zhang N, Patra BN, La Cruz De M, Ray A, et al. Polyglutamine disease modeling: epitope based screen for homologous recombination using CRISPR/Cas9 system. *PLoS Curr.* 2014;15:6.
101. Rodriguez-Perales S, Cano F, Lobato MN, Rabbitts TH. MLL gene fusions in human leukaemias: in vivo modelling to recapitulate these primary tumourigenic events. *Int J Hematol.* 2008;87(1):3–9.
102. Maresch R, Mueller S, Veltkamp C, Öllinger R, Friedrich M, Heid I, et al. Multiplexed pancreatic genome engineering and cancer induction by transfection-based CRISPR/Cas9 delivery in mice. *Nat Commun.* 2016;7:10770.
103. Sánchez-Rivera FJ, Papagiannakopoulos T, Romero R, Tammela T, Bauer MR, Bhutkar A, et al. Rapid modelling of cooperating genetic events in cancer through somatic genome editing. *Nature.* 2014;516(7531):428–431.
104. Mao X-Y, Dai J-X, Zhou H-H, Liu Z-Q, Jin W-L. Brain tumor modeling using the CRISPR/Cas9 system: state of the art and view to the future. *Oncotarget.* 2016;7(22):33461–33471.
105. Torres-Ruiz R, Martinez-Lage M, Martin MC, Garcia A, Bueno C, Castaño J, et al. Efficient recreation of t(11;22) EWSR1-FLI1(+) in human stem cells using CRISPR/Cas9. *Stem Cell Rep.* 2017;8(5):1408–1420.
106. Matano M, Date S, Shimokawa M, Takano A, Fujii M, Ohta Y, et al. Modeling colorectal cancer using CRISPR-Cas9-mediated engineering of human intestinal organoids. *Nat Med.* 2015;21(3):256–262.
107. Zhang M, D'Aniello C, Verkerk AO, Wrobel E, Frank S, Ward-van Oostwaard D, et al. Recessive cardiac phenotypes in induced pluripotent stem cell models of Jervell and Lange-Nielsen syndrome: disease mechanisms and pharmacological rescue. *Proc Natl Acad Sci USA.* 2014;111(50):E5383–E5392.
108. Wang Y, Liang P, Lan F, Wu H, Lisowski L, Gu M, et al. Genome editing of isogenic human induced pluripotent stem cells recapitulates long QT phenotype for drug testing. *J Am Coll Cardiol.* 2014;64(5):451–459.
109. Gupta RM, Meissner TB, Cowan CA, Musunuru K. Genome-edited human pluripotent stem cell-derived macrophages as a model of reverse cholesterol transport—brief report. *ArteriosclerThrombVascBiol.* 2016;36(1):15–18.
110. Wang G, McCain ML, Yang L, He A, Pasqualini FS, Agarwal A, et al. Modeling the mitochondrial cardiomyopathy of Barth syndrome with induced pluripotent stem cell and heart-on-chip technologies. *Nat Med.* 2014;20(6):616–623.
111. Carroll KJ, Makarewich CA, McAnally J, Anderson DM, Zentilin L, Liu N, et al. A mouse model for adult cardiac-specific gene deletion with CRISPR/Cas9. *Proc Natl Acad Sci USA.* 2016;113(2):338–343.
112. Ding Q, Strong A, Patel KM, Ng S-L, Gosis BS, Regan SN, et al. Permanent alteration of PCSK9 with in vivo CRISPR-Cas9 genome editing. *Circ Res.* 2014;115(5):488–492.
113. Liao H-K, Gu Y, Diaz A, Marlett J, Takahashi Y, Li M, et al. Use of the CRISPR/Cas9 system as an intracellular defense against HIV-1 infection in human cells. *Nat Commun.* 2015;6:6413.

114. Yin C, Zhang T, Qu X, Zhang Y, Putatunda R, Xiao X, et al. In vivo excision of HIV-1 provirus by saCas9 and multiplex single-guide RNAs in animal models. *Mol Ther.* 2017;25(5):1168–1186.
115. Ramanan V, Shlomai A, Cox DBT, Schwartz RE, Michailidis E, Bhatta A, et al. CRISPR/Cas9 cleavage of viral DNA efficiently suppresses hepatitis B virus. *Sci Rep.* 2015;5:10833.
116. Li Z, Bi Y, Xiao H, Sun L, Ren Y, Li Y, et al. CRISPR-Cas9 system-driven site-specific selection pressure on *Herpes simplex virus* genomes. *Virus Res.* 2017.
117. Zhou J, Shen B, Zhang W, Wang J, Yang J, Chen L, et al. One-step generation of different immunodeficient mice with multiple gene modifications by CRISPR/Cas9 mediated genome engineering. *Int J Biochem Cell Biol.* 2014;46:49–55.
118. Hoodless LJ, Lucas CD, Duffin R, Denvir MA, Haslett C, Tucker CS, et al. Genetic and pharmacological inhibition of CDK9 drives neutrophil apoptosis to resolve inflammation in zebrafish in vivo. *Sci Rep.* 2016;5(1):36980.
119. Hurlbut JB. Limits of responsibility: genome editing, asilomar, and the politics of deliberation. *Hastings Cent Rep.* 2015;45(5):11–14.
120. Keener AB. Delivering the goods: scientists seek a way to make CRISPR-Cas gene editing more targeted. *Nat Med.* 2015;21:1239–1241.
121. Chan S, Donovan PJ, Douglas T, Gyngell C, Harris J, Lovell-Badge R, et al. Genome editing technologies and human germline genetic modification: The Hinxton Group Consensus Statement. *Am J Bioeth.* 2015;15(12):42–47.

CHAPTER THREE

Dynamics of Indel Profiles Induced by Various CRISPR/Cas9 Delivery Methods

Michael Kosicki*,[a], Sandeep S. Rajan*,[a], Flaminia C. Lorenzetti†, Hans H. Wandall†, Yoshiki Narimatsu†,‡, Emmanouil Metzakopian*,[1], Eric P. Bennett†,[1]

* Wellcome Trust Sanger Institute, Hinxton, Cambridge, United Kingdom
† Departments of Cellular and Molecular Medicine and Odontology, Copenhagen Center for Glycomics, University of Copenhagen, Copenhagen, Denmark
‡ GlycoDisplay Aps, Copenhagen, Denmark
[1] Corresponding authors. E-mail address: em11@sanger.ac.uk; epb@sund.ku.dk
[a] These authors have contributed equally to the work presented.

Contents

Abstract

The introduction of CRISPR/Cas9 gene editing in mammalian cells is a scientific breakthrough, which has greatly affected basic research and gene therapy. The

Progress in Molecular Biology and Translational Science, Volume 152
ISSN 1877-1173
http://dx.doi.org/10.1016/bs.pmbts.2017.09.003

simplicity and general access to CRISPR/Cas9 reagents has in an unprecedented manner "democratized" gene targeting in biomedical research, enabling genetic engineering of any gene in any cell, tissue, organ, and organism. The ability for fast, precise, and efficient profiling of the double-stranded break induced insertions and deletions (indels), mediated by any of the available programmable nucleases, is paramount to any given gene targeting approach. In this study we review the most commonly used indel detection methods and using a robust, sensitive, and cost efficient Indel Detection by Amplicon Analysis method, we have investigated the impact of the most commonly used CRISPR/Cas9 delivery formats, including lentivirus transduction, plasmid lipofection, and ribo nuclear protein electroporation, on the dynamics of indel profile formation. We observe rapid indel formation using RNP electroporation, especially with synthetic stabilized gRNA, as well as long-term decline in overall indel frequency with lipofectamine-based, plasmid transfection methods. Most methods reach peak editing on day 2–3 postdelivery. Furthermore, we find relative increase in frequency of larger size indels (>6 bp) under condition of persistent editing using stably integrated lentiviral gRNA and Cas9 vectors.

1. INTRODUCTION

1.1 Genome Editing and DNA Repair

Genome editing tools, such as meganucleases,[1] ZFN,[2] TALENs,[3] and CRISPR/Cas9,[4,5] work by introducing a double-stranded break (DSB) at a defined genomic position. The DSB is resolved by cellular DNA repair pathways, which either results in a mutation, incorporation of a donor template (if provided) or restoration of the original allele. In the latter case, the cycle is repeated until the recognition site is destroyed or until the nuclease activity is terminated.

Various pathways are involved in the resolution of DSBs, such as non-homologous end joining (NHEJ),[6] microhomology mediated end joining (MMEJ),[7] and homologous recombination (HR).[8,9] In brief, NHEJ in normal situations result in perfect repair, but upon excessive insult, such as extensive nuclease cutting causes formation of indels in the original allele, and involves multiple steps such as stabilization, resection, and ligation of the broken ends. MMEJ, also known as alternative end joining,[10] has been discovered as a "salvage" pathway, which becomes more prominent in cells deficient for NHEJ. In MMEJ, the broken ends of the DNA are resected until a microhomology (>3 bp) is found and the ends can be rejoined. This often results in larger deletions (>3 bp) and occasionally structural variants, such as translocations.[11] HR shares the initial end resection step with

MMEJs,[12] but mediates a seamless repair of the break. It achieves that by using the second copy of the chromosome (sister chromatid), which is present in the S/G2 phase of the cell cycle, as a template for repair.[6,13] Integration of exogenous templates that may be implemented for a specific DNA modification (genome editing), is also mediated by HR.

We and others have shown that the indels generated by one of the genome editing nucleases, Cas9, are specific to a given guide RNA design, nonrandomly induced, but independent of the genomic context and cell type.[14–20] Activity of different DNA repair pathways likely contributes to that profile, as demonstrated by relative depletion of short (<3 bp) indels by inhibition of NHEJ.[16] However, the indel profiles generated by different gRNA/Cas9 delivery methods were not systematically studied, motivating our investigation in this chapter.

1.2 Indel Detection Methodologies

The most common goal of genome editing experiments is site-specific mutagenesis of protein coding genes. A DSB introduced in an exon may resolve as an out-of-frame indel, which in turn often results in loss of protein expression. When selection for this outcome is impossible or impractical (e.g., in vivo), maximizing the efficiency of deleterious indel generation becomes paramount. Computational prediction of this efficiency can improve the performance over large sets of targets (>10,000), but still fails at individual target level. Therefore, for low to medium throughput experiments (~1–1000) direct validation of genome editing by detection of indels is warranted. Of notice, a good indel generation efficiency may correlate with strong affinity of the nuclease for its target and thus serve as a proxy for efficiency of other targetable activity, such as gene transactivation or inhibition.

Methods for detection of indels differ widely in their ability to resolve alleles, sensitivity, requirement for technical expertise, difficulty of results analysis, cost, time and effort required. On one end of the spectrum, microarrays combined with bioinformatical imputation allow genotyping of millions of Single Nucleotide Polymorphisms (SNPs) and structural variants in hundreds of thousands of patients, at a very low-cost per allele. On the other end, individual researchers genotype cross-bred animals using simple discriminatory PCR and resolve the sequence of individual alleles in pure cell lines by Sanger sequencing.

In this study, we briefly review the most commonly used indel detection methodologies and focus on the middle ground of allele detection, which is assessment of genome editing experiments conducted on a pool of cells.

We identify Enzyme Mismatch Cleavage (EMC),[21] Tracking of Indels by DEcomposition (TIDE),[22] Next Generation Sequencing (NGS), and Indel Detection by Amplicon Analysis (IDAA)[15] as the most fitting methods for this purpose (Fig. 1). Each method requires PCR amplification of the genomic region of interest, followed by varying methods of allele separation and detection.

EMC, colloquially known as T7 or Surveyor assay, is perhaps the most commonly used method to demonstrate gene editing, owing to its low cost, ease of use, and rapid turnover time. In EMC, a pool of genomic PCR products stemming from differently edited alleles is denatured and rehybridized, resulting in formation of homo- and heteroduplexes. The latter are selectively digested by one of the heteroduplex sensitive endonucleases, such as T4 endonuclease VII (T4E7), endonuclease V (EndoV), T7 endonuclease I (T7EI), CELI, or Surveyor. Resolution of the wild-type sized homoduplexes and digested heteroduplex is achieved by agarose gel electrophoresis. It is followed by simple visual assessment or quantification of band intensities to determine an efficiency score. The entire procedure can easily be completed on a single day. However, EMC suffers from low sensitivity, resulting partly from the detection methods and partly due to aberrant cleavage of some genomic structures. In particular, T7EI and Surveyor possess a strong nonspecific activity and display a preference for heteroduplex DNA formed by deletions rather than single point mutations.[23] This is particularly problematic in CRISPR/Cas9-based gene editing experiments, which are

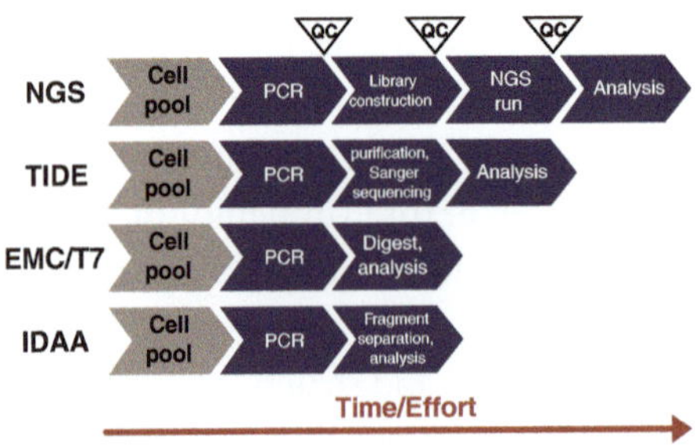

	Sensitivity	Resolution	Scalability	Cost
NGS	**High**	**High**	Low	High
TIDE	Medium	Medium	Medium	Medium
EMC/T7	Low	Low	**High**	**Low**
IDAA	**High**	Medium	**High**	**Low**

Fig. 1 Methods for analysis of genome editing in cell pools. Relative time/effort required (axis shown below) for completing the necessary steps for next generation sequencing (NGS), indel detection by amplicon analysis (IDAA), tracking of indels by decomposition (TIDE), and enzyme mismatch cleavage (EMC). QC represents necessary quality control steps required for NGS. Table scores the performance (low, medium, high) of the respective methods with regard to sensitivity, resolution, scalability, and cost. Preferred choices are shown in bold font.

known to predominantly generate minor indels.[15,16,18] If the original locus is heterozygous, it may be intractable to EMC analysis altogether due to heteroduplex formation in the wild type. Furthermore, EMC does not provide information on the nature of detected indels.

TIDE, surpasses EMC in terms of sensitivity and allelic resolution at a slightly increased cost in time and money. In TIDE, PCR products are Sanger sequenced and the resulting traces are fed into an online analysis software (https://tide.nki.nl/), which deconvolutes the trace mixture. The result is a discrete profile of indels. As wild-type sequence is used as a control, heterozygous loci should in theory be amenable to TIDE analysis. While TIDE's sensitivity is good (1%–2%), it will miss rare alleles. Furthermore, no information about the sequence of alleles is given (beyond the identity of +1 insertion). Alleles of the same size, but different sequence identity, are thus clustered together. TIDE is designed specifically for CRISPR/Cas9 experiments, though this may potentially be circumvented by providing a "fake" gRNA sequence. As the original code for the tool is only available on request, the casual user may be dependent on continued maintenance of the online service.

NGS, at the top end of the spectrum, provides very-high sensitivity and unparalleled allelic resolution, at a much higher cost in time, effort and money (except in cases of high throughput read-outs). In a typical NGS experiment, the region of interest (100–500 bp) is amplified and barcoded. Samples from different experiments are pooled together, adapter ligated and sequenced. Bioinformatical analysis may be either be conducted using "in-house" scripts, online tools, such as CRISPR Genome Analyzer,[24] CRISPResso,[25] or packages, such as CRISP-R (https://bioconductor.org/packages/release/bioc/html/CrispRVariants.html). NGS reveal the exact sequence identity of the alleles at about 0.1% sensitivity. Despite the availability of online tools and commercial services, NGS remains a relatively slow, complex and costly option, especially at the middle-throughput scale of genome editing experiments.

1.3 IDAA for Indel Detection and gRNA Selection

We have developed IDAA—a simple method for allele detection characterized by high sensitivity, low cost, as well as good resolution and turnover time. In IDAA, the genomic PCR products are fluorescently tagged by addition of a third, universal primer (Fig. 2B). The amplified fluorescent products are resolved and quantified at a basepair resolution, using a fragment analyzer machine (similar to those used in Sanger sequencing), yielding an indel profile of the targeted region (Fig. 2C; see Lonowski[18] for details).

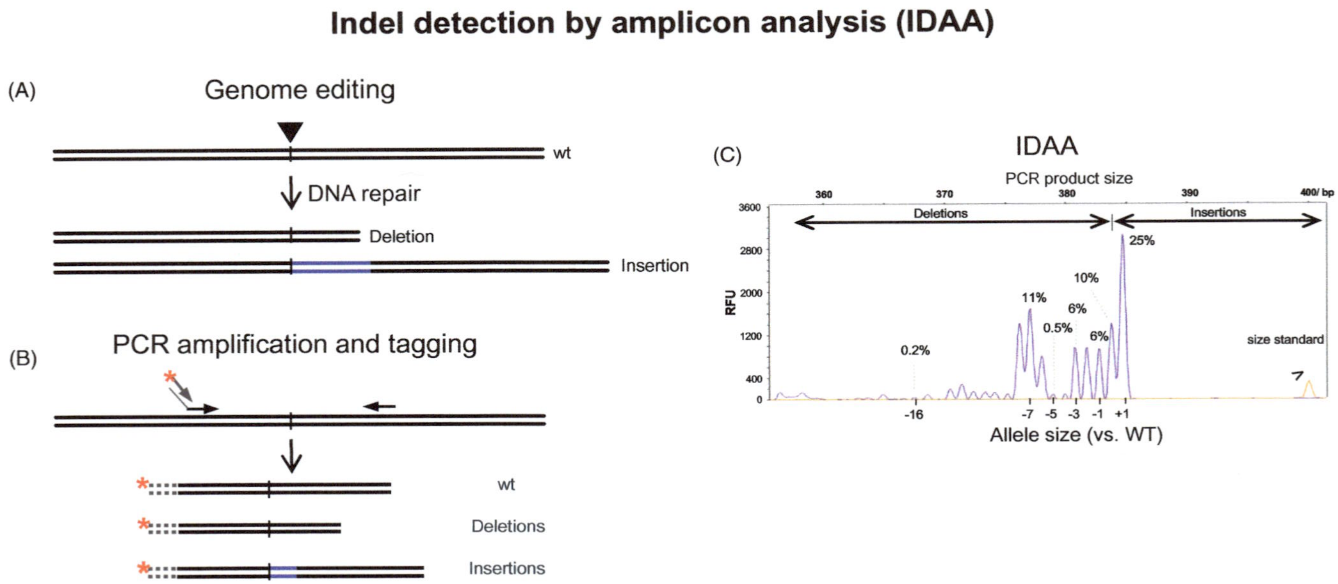

Fig. 2 IDAA workflow. *Panel (A)*: Cell pool genome editing and repair. A nuclease programed to cut the region of choice introduces a double-stranded-break, which is resolved by host DNA repair. This may results in a mutation (deletion or insertion), if nonhomologous end-joining pathway is employed. *Panel (B)*: Fluorophore labeling of target specific amplicons. The region of interest is amplified using a triprimer PCR reaction. One of the two genomic primers contains an overhang that allows annealing and amplification by a universal fluorescently labeled third primer. *Panel (C)*: Allele resolution and quantification. Fluorescent PCR products are separated at single basepair resolution and quantified using a fragment analysis machine (a standard DNA sequenator). PCR product size (top *x*-axis) can be expressed as the allele size versus wild-type (wt, bottom *x*-axis) and the amount of given product/allele corresponds to the intensity of the peak expressed in relative fluorescent units (RFU, *y*-axis). Standard size marker (orange) allows absolute quantification and size assignment. *Figure has been adapted from Lonowski LA, Narimatsu Y, Riaz A, et al. Genome editing using FACS enrichment of nuclease-expressing cells and indel detection by amplicon analysis. Nat Protoc. 2012;12(3):581–603.*[18]

In comparison to the other methods described, IDAA parallels TIDE in allelic resolution, while being more scalable, as cost efficient as EMC and as sensitive as NGS (capable of detecting picogram amounts of PCR products), making it a preferred method for indel detection for any genome editing experiment.

In recent studies we have revealed that the resolution and sensitivity of IDAA is similar to NGS and compared EMC with IDAA after delivery of CRISPR/Cas9 reagents in cell pools. These studies corroborates previous findings of the high reproducibility of IDAA, superior allelic resolution and comparable indel profile with NGS obtained profiles on matched samples.[18,19] Importantly, we show that indel profile and sequence of alleles induced by a given gRNA design is reproducible and gives rise to nonrandom indels.[19]

Using IDAA, we have experimentally validated more than a thousand of different gRNA designs in various model cancer cell lines, including HEK293, K562, CHO-K1, and U2OS, and found that at least half of the designs caused little or no indel formation. Therefore, we have set a threshold for gRNA designs we designate as "validated gRNA designs" to >30% cutting efficiency (indel formation) in HEK293, CHO, or K562 cells (manuscript in preparation).

Here, we expand on our previous findings and report that indel profiles are reliably detected in small amounts of samples (as few as 250 cells), with stochastic allele detection possible in as few as 10 cells (Fig. 3). We speculate that single cell detection of alleles can be achieved, if proper amplification is employed.

2. MATERIALS AND METHODS

2.1 gRNAs, Plasmid Vectors and Cell Lines

We targeted *ST6GALNAC1*, encoding a glycosyltransferase involved in posttranslational modification, which is silent in many normal human cells and tissues including the human cell line HEK293 used in this study.[26] The gRNA for *ST6GALNAC1* (5′-ACGGTGTCAGAGAAGCACCAggg-3′, PAM in lowercase) was chosen for its high efficiency from our library targeting the glycogenome.[27]

All plasmid vectors used in the study share a common backbone containing PiggyBac inverted repeats and lentiviral terminal repeats.[28] The gRNA

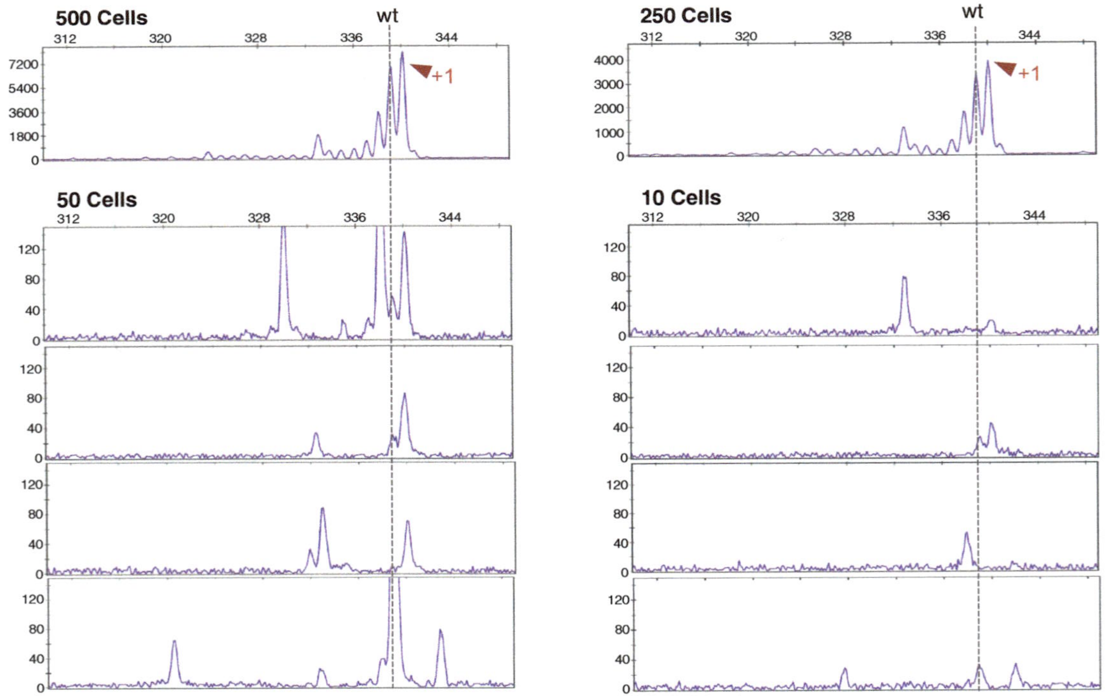

Fig. 3 IDAA sensitivity. HEK293 cell line stably expressing Cas9 was transfected with a plasmid encoding a validated *GALNT3* gRNA expression cassette (5′-TATGGAAGTAACCATAACCG-3′; manuscript in preparation). Three days post transfection cells were counted and crude QuickExtract (Epicentre, Madison, USA) lysate from 10^4 cells diluted down to amounts representing the indicated number of cells to be analyzed by IDAA. Four independent replicate samples for 50 and 10 cells were analyzed by IDAA according to previously published guidelines.[18] Representative fluorograms from duplicate experiments set up are shown. Characteristic +1 bp major indel and wild type unmodified alleles (*stippled line*) are indicated. Should be noted that with the amplification conditions used, for the lower cell number analysis (10 cells), indels detected are below the recommended IDAA detection threshold of 150RFU.

expressing plasmid contains a U6 promoter and PGK-BFP-2A-puro cassette. The Cas9 expressing plasmids contains EFS-Cas9-2A-blast cassette.[29] For the RNP experiments, regular or stabilized synthetic SygRNA™ crRNA and tracrRNA (TRACRRNA05N or TracrRNA05N, Merck KGaA, Darmstadt, Germany) were hybridized in IDT Duplex Buffer by heating to 95°C for 2 min and cooling down to 25°C at the rate of 0.1°C/s. In stabilized gRNAs, three terminal ribonucleotides from both 5′ and 3′ end were 2′O-methylated and their phosphodiester linkages were replaced with phosphorothioate ones. Cas9 protein was obtained from NEB (EnGen-Cas9-NLS, #M0646).

Single-cell cloned, MAP tested, human HEK293 cells were used in all experiments. For the lentiviral condition, HEK293 cells were transduced with Cas9 lentivirus produced in 293T cells, selected using 10 μg/mL blasticidin and single cell cloned. These cells were under continuous blasticidin selection throughout the experiment. Where indicated, puromycin was used for selection at 3 μg/mL. Cells were maintained in DMEM supplied with L-glutamate, β-mercaptoethanol, and 15% fetal calf serum, without any antimicrobial agents.

2.2 Cas9 and gRNA Delivery Conditions

For the RNP-electroporation condition, 1,50,000 HEK cells were electroporated using Neon (Invitrogen) system (1700 V, 20 ms, 1 pulse) with 10 pmol Cas9 protein and 10 pmol hybridized gRNA. For other methods, HEK cells were seeded in 24 W plates the day before the transfection so as to achieve 50%–70% confluency. For PiggyBac and transient plasmid conditions, cells were transfected with 150 ng gRNA plasmid, 150 ng Cas9 plasmid, and 50 ng of either hyperactive PiggyBac transposase[30] or carrier plasmid (pBluescript II SK+). For the protein + plasmid (P&P) and protein + plasmid + carrier (P&P-carrier) conditions, cells were transfected with two separately prepared mixes: (1) 3 pmol Cas9 protein, (2) 150 ng gRNA plasmid with or without 200 ng carrier plasmid (pBluescript II SK+). For RNP-lipofectamine conditions, cells were transfected with 3 pmol Cas9 protein, 3 pmol hybridized gRNA (regular or stabilized). Plus reagent was added at 1 μL per 1 μg plasmid and Lipofectamine 3000 at double that volume. Cas9 protein was mixed with 1.5 μL Lipofectamine 3000.

2.3 Efficiency Measurements

Percentage of BFP positive cells was assessed using Cytoflex or BD Fortessa flow cytometers. In addition to IDAA method, some samples were assessed for cutting efficiency using TIDE. Same PCR amplicons, produced using

AmpliTaq Gold (ABI/Life Technologies, USA) and ST6EXT/REV primers (5′-*AGCTGACCGGCAGCAAAATTG* TGGAAGAGCCCAGAAAAA-GA-3′/5′-TCCTTAGGTGGATGACTGC-3′) (extension overhang for IDAA underlined) were utilized for both methods. IDAA protocol is described elsewhere.[18] Analysis of data from IDAA experiments was performed in R using "binner" package (https://github.com/plantarum/binner). Visualizations were made using "ggplot" package. Efficiency score was calculated as 1—(wild-type peak intensity/sum of wild-type and prominent peaks intensities). Spurious "−1" indel signal present in wild-type samples was estimated to be around 10% of wild-type peak. This intensity was subtracted from "−1" peak and added to wild-type peak in all samples. Prominent peaks were defined using an arbitrary cutoff on the sum of intensities over many experiments and included −22, −14, −13, −11, −9, −8, −6, −4, −3, −2, −1, and +1 peaks.

3. RESULTS

3.1 Rationale and Design of the Study

The Cas9 protein and guide RNA may be delivered into cells in a variety of forms (e.g., plasmid DNA, mRNA, protein, lentivirus) and using a variety of methods (e.g., electroporation, lipofection, calcium phosphate transfection, transduction). We wondered how the form and the method of Cas9/gRNA delivery influences indel formation, especially with regards to relative indel composition and overall dynamics of the process. To study this, we collected cells at multiple timepoints postdelivery and analyzed the indels using IDAA. We chose to target a commonly used HEK293 cell line, known to be amenable to many delivery methods and shown to achieve high editing rates. A validated, highly efficient guide RNA against *ST6GALNAC1*, which is silent in HEK cells and was picked in hope of avoiding knockout-specific proliferative effects. In addition to cargo delivery methods broadly used for gene-editing, which includes RNP electroporation, RNP lipofection, plasmid gRNA, and Cas9 lipofection and lentiviral gRNA transduction into cells stably expressing Cas9, we also included PiggyBac transposition and co-lipofection of plasmid gRNA with protein Cas9 (Fig. 4). We took advantage of a plasmid backbone developed in our group, which contains both lentiviral and PiggyBac functional elements, to minimize differences between methods.[28]

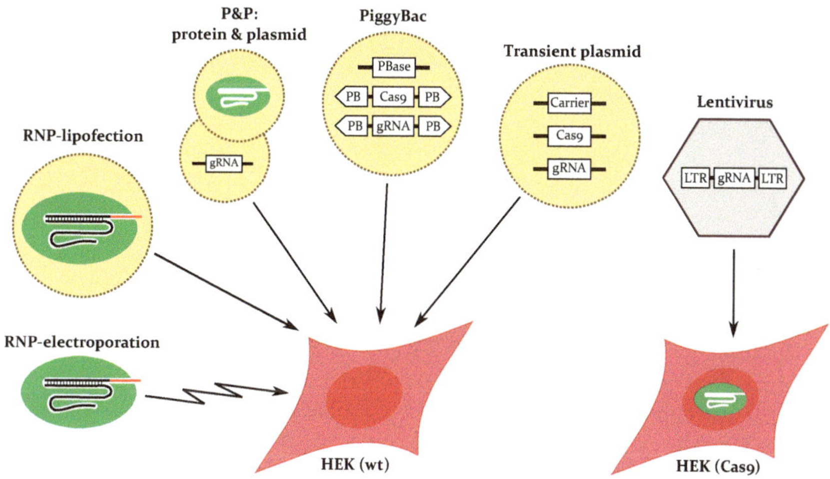

Fig. 4 Cas9 and gRNA delivery methods. Cas9 and gRNA were delivered to HEK cells in a variety of ways. The same gRNA and Cas9 plasmids were used for all experiments. *Yellow circle* indicates transfection with Lipofectamine 3000. RNP-electro: electroporation of Cas9 protein and synthetic gRNA using Neon Transfection System. RNP-lipo—as RNP-electro, using Lipofectamine 3000. P&P: protein & plasmid—transfection of Cas9 protein and gRNA-encoding plasmid. PB: PiggyBac—transfection of plasmids encoding Cas9 and gRNA together with PiggyBac transposase (resulting in stable cellular integration). trans: transient—as PB, with transposase replaced by a carrier plasmid. lenti: lentiviral—transduction of HEK cells stably expressing Cas9 with gRNA lentivirus.

3.2 Transfection and Integration Dynamics

Methods compared in this study result in either stable (lentivirus, PiggyBac) or transient (P&P, RNP, transient plasmid) expression of Cas9 and gRNA. As the Cas9 plasmid contains no fluorescent marker, we monitored the BFP expression from the BFP-2A-puro cassette in the gRNA plasmid as a proxy for the overall transfection and integration efficiency (Fig. 5A).

PiggyBac and transient methods resulted in the highest transfection efficiencies on day 2 (>70%), followed by P&P-carrier and P&P (55% and 40%; Fig. 5B). As expected, the BFP expression was all but extinguished in the transient, P&P and P&P-carrier conditions by day 14. In contrast, 20% of the cells in the PiggyBac condition (about 1/4 of all transfected cells on day 2) remained BFP positive at this time, indicating stable transposition. Short-term selection (days 10–14) for the gRNA plasmid increased the percentage of positive cells to ~44%. Interestingly, a small number of BFP positive cells in the transient condition at this time may indicate creation of transgenic cell lines.

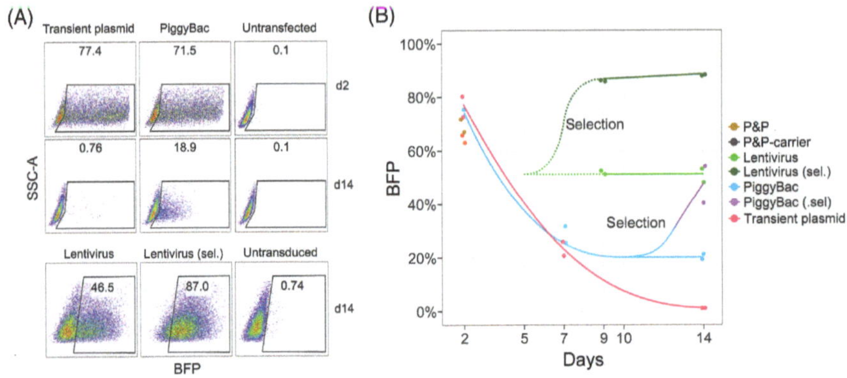

Fig. 5 Transfection efficiency over time. HEK cells transfected with Cas9 and gRNA were analyzed by flow cytometry at various time points. BFP fluorescence comes from gRNA expressing plasmids and thus likely overestimates the actual percentage of Cas9/gRNA double-positive cells. The exception is lentiviral transduction, where cells constitutively express Cas9. As RNP has no fluorescent component, the transfection efficiency of this methods is unknown. lenti.: lentiviral, sel.: selected for gRNA construct. *Panel (A)*: Examples of flow cytometry plots. Samples were assessed using Cytoflex machine, except for the lentiviral method, which was assessed using BD Fortessa. *Panel (B)*: Comparison of transfection levels across methods and timepoints. *Solid line* indicates the most likely path between collected timepoints. *Dotted line* connects day 7, the time when selection of lentiviral condition was started (no data available, efficiency inferred) to day 9 for selected and unselected samples. Horizontal jitter was added for clarity.

While data from early timepoints in lentiviral transductions is missing, the percentage of BFP positive cells was maintained at approximately 50% between days 9 and 14 posttransduction, indicating stable integration. Selection for puro resistance between days 5 and 14 significantly increased this proportion. Cells that fall outside of the gate established on the negative control sample, may indicate either an incomplete selection or the fact some resistant cells express BFP at a lower level. As neither component of the RNP was fluorescent, we could not monitor the transfection efficiency of this method.

3.3 Efficiency of Generating Indels Over Time

We studied the dynamics of indel generation using IDAA. For RNP and P&P methods we collected samples between 6 h and 3 days, on the assumption that Cas9 protein is degraded by that time. For other methods, we continued collecting samples up till day 14 (examples of IDAA indel profiles in Fig. 6A). The indel generation efficiency was calculated, for a given

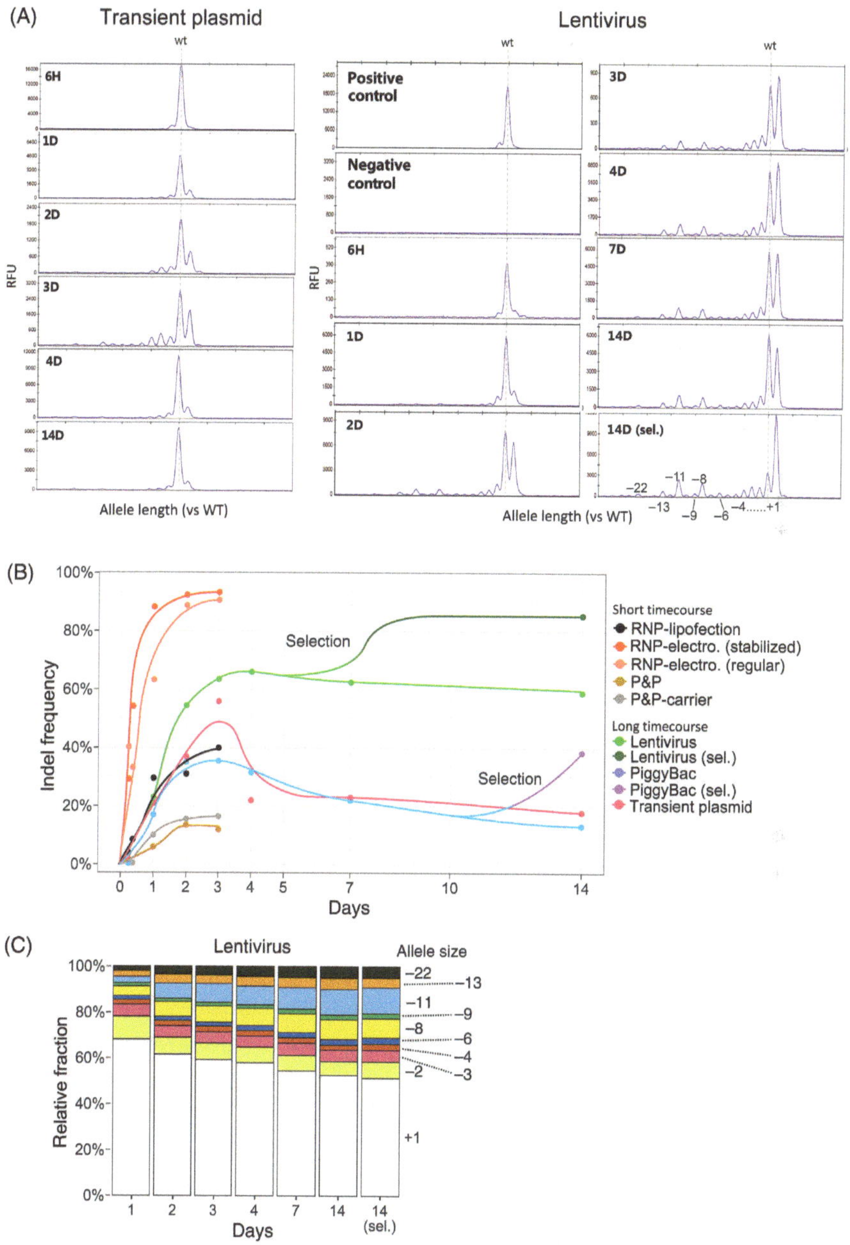

Fig. 6 Indel frequency and profiles over time. HEK cells transfected with Cas9 and gRNA were analyzed by IDAA at various timepoints. RNP samples were only collected up to day 3. *Panel (A)*: Examples of IDAA plots for lentivirus and transient plasmid conditions. electro.: electroporation, sel.: selected for gRNA construct. Prominent peaks used for efficiency calculation are indicated in the selected sample. *Asterisk* indicates truncated

sample, as a ratio of intensity of prominent, nonwild-type peaks to all the peaks (Fig. 6B). As the wild-type sized peak may represent rare balanced indels or SNPs in addition to the wild-type allele, this efficiency may be slightly underestimated.

On day 3 posttransfection, RNP-electroporation was the most efficient indel inducing method in this study (91%–93%), followed by lentivirus (64%), transient plasmid (56%), RNP-lipofection, PiggyBac (36%–40%), and P&P/P&P-carrier (12%–17%). The transfection efficiency broadly correlated with the indel generation efficiency, except for P&P and P&P-carrier, which induced relatively lower levels of indels. Since Lipofectamine 3000 complexes of Cas9 protein and gRNA plasmid in these methods were prepared separately, it is possible the rate of cotransfection was much lower than that of individual components. For transient plasmid transfection, no smooth curve fit could be found for experimental data, indicating some timepoints were outliers. Interestingly, indel frequency was slightly lower in transient plasmid condition than in PiggyBac on day 14, despite stable PiggyBac integration. However, selection for stable PiggyBac integrants between days 10 and 14 near doubled the final percentage of edited alleles (17%–38%).

Most of the methods of cargo delivery reached their maximum efficiency on day 2 and 3. Both RNP-electroporation and RNP-lipofection show earliest indels formation at 6 and 9 h posttransfection. Interestingly, transient plasmid and PiggyBac exhibited a significant decrease in efficiency after day 3, while lentivirus plateaued. This may indicate an anti-proliferative effect of transfection.

3.4 Indel Profiles Over Time

To investigate whether the allelic composition stays stable or fluctuates over time and across methods we quantified the relative abundances of indel peaks. As larger deletion indels (e.g., −22, −13, −11, and −8) may correspond to microhomologies that we found around the target site (data not shown), we wondered if they would appear later than the smaller ones. This would be consistent with the published observation that MMEJ is a slower repair

◀ product, which is subtracted computationally for efficiency calculation and is likely the result of a failure to add A-tail by Taq polymerase. Representative panels from duplicate experiments are shown. *Panel (B)*: Indel frequency over time. *Dots* represent single samples or an average of two replicates. *Solid line* indicates most likely path between collected timepoints. *Panel (C)*: Indel profiles over time in the lentiviral sample. Prominent peaks, as indicated in Panel (A), are colored.

pathways than NHEJ,[31] which usually creates small indels. The indel profile in lentiviral condition, which yielded most reliable data, indicates a tendency towards larger indel formation over time (Fig. 6C). This suggests either some large indels are formed later in the timecourse or that small indels (such as +1 insertion) and single nucleotide variants[19] are susceptible to recutting, yielding larger indels upon mutagenic repair (as previously suggested[19]).

3.5 Dynamic Effect of gRNA Stability on Indel Efficiency

It has been proposed that gRNAs protected from cellular exonucleases by chemical modifications may increase the overall indel generation. Therefore, we studied the dynamics of indel generation using IDAA in cells electroporated with Cas9 protein and either regular or stabilized gRNAs. Both gRNAs showed similar, rapid indel formation by sixth hour postelectroporation and demonstrate comparable indel profile and maximum efficiency at the end of the experiment on day 3 (Figs. 6B and 7). However, the stabilized gRNA reached its maximum efficiency on day 1, faster than the regular gRNA by about 24 h. It is known, that guide RNA loading is a key regulator of Cas9 enzyme function[32] and we speculate that stabilized gRNA provides for superior binding of the RNP complex to its target sequence or that the conformational Cas9 changes that take place upon stabilized gRNA binding improve positioning of the catalytic RuvC and HNH domains for optimal nuclease activity.

3.6 Comparison of IDAA and TIDE Methods

As both IDAA and TIDE generate indel profiles, we decided to compare the results for selected samples. While the overall profiles were similar, we have noticed that sensitivity of TIDE differed depending on which primer was used for the Sanger sequencing reaction (Fig. 8). As IDAA does not involve a sequencing step, it is not susceptible to this problem.

4. CONCLUDING REMARKS

In this study we briefly reviewed the application of the most commonly used indel detection methods: EMC, TIDE, NGS, and IDAA. Based on previous comparative studies of EMC, NGS, and IDAA,[18,19] and attributes of IDAA, to the best of our knowledge, this study for the first time has undertaken an in depth analysis of how different CRISPR/Cas9 formats

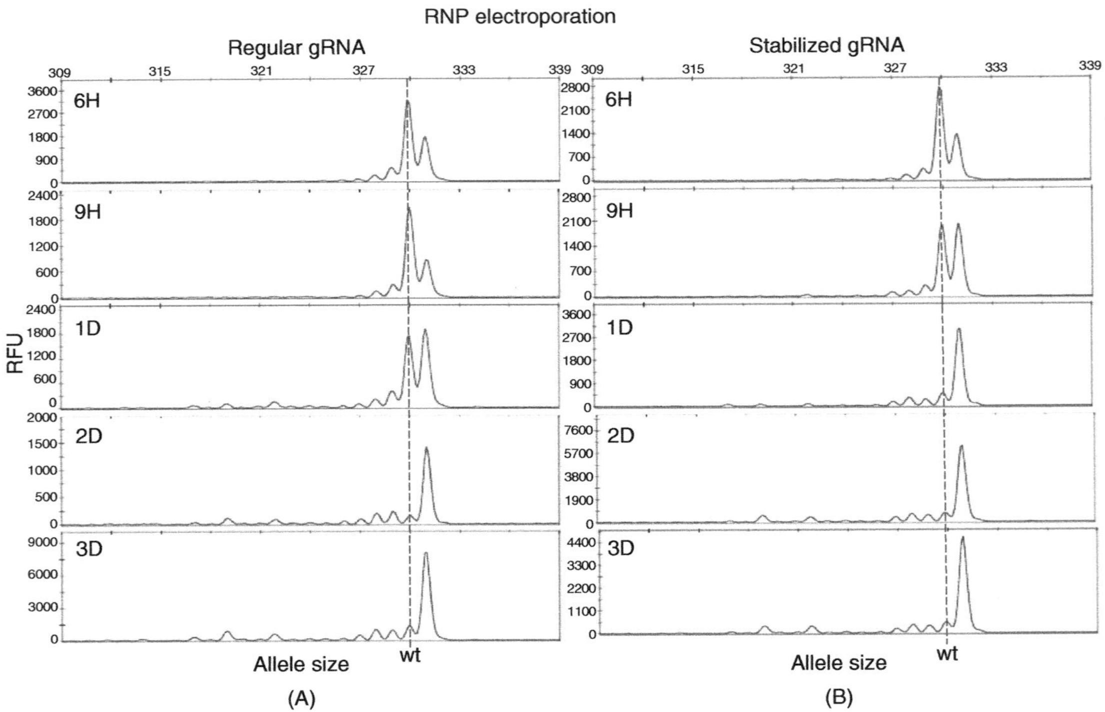

Fig. 7 Comparison of normal and stabilized synthetic gRNAs. HEK cells were electroporated using neon transfection system with protein Cas9 and either regular or stabilized synthetic gRNAs. Panels (A) and (B) show the dynamics of indels appearing shortly after cellular delivery (6 h) and up to 3 days post delivery. Representative panels from duplicate experiments are shown.

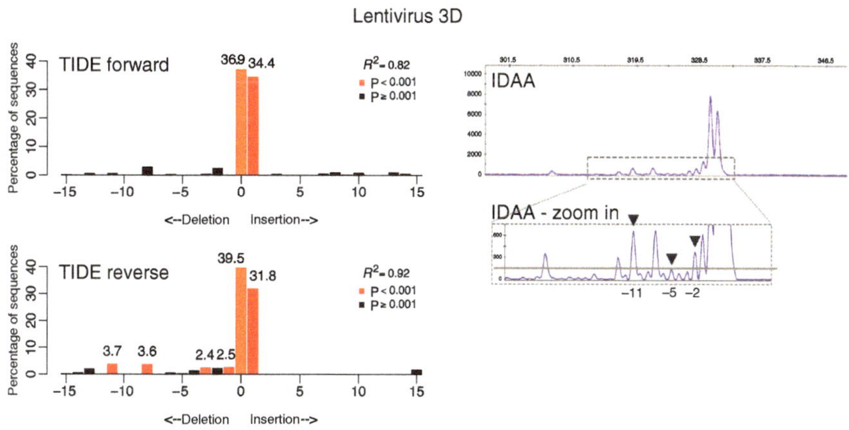

Fig. 8 Comparison of IDAA and TIDE. The sample shown in Fig. 6, Panel A, lentivirus 3D was indel profiled by TIDE analysis using both the forward or reverse Sanger sequencing raw data files. The results were compared to IDAA generated profiles, where a zoom in of the indel peaks is provided as inset for clarity and recommended IDAA detection threshold (150RFU) indicated by the horizontal line. Selected low-abundant indels are indicated with *black triangles*.

impact on indel profile and dynamics in the immediate time period after cellular delivery. Commonly used CRISPR/Cas9 formats including: lentiviral, PiggyBac, RNP and plasmid were delivered to cells using transduction, lipofection, or electroporation. Of the delivery methods tested, indels are most rapidly induced, notably within 6 h, following RNP delivery by electroporation and importantly, when precomplexed with protected synthetic gRNA's that give rise to indel cleavage efficiencies of 90% within 24 h. Similar efficacies could also be obtained by lentiviral transduction, but only after prolonged cell culture for 14 days and using antibiotic selection.

Taken together, our results demonstrate that cellular repair of CRISPR/Cas9 induced DSBs in the human HEK293 cell line model, gives rise to indel profiles that are independent of the delivery format used and remain constant over time after delivery. Thus, it remains to be determined how these observations translate into use of CRISPR/Cas9 delivery formats in primary and stem cells, tissues, and organs.

ACKNOWLEDGMENTS

We thank Camilla Andersen (Copenhagen Center for Glycomics, Department of Odontology, University of Copenhagen) for excellent technical assistance. We also thank Gurpreet Balrey from Merck KGaA, Darmstadt, Germany for supplying the investigators with the SygRNA™ synthetic nucleic acid reagents used.

University of Copenhagen Excellence Programme for Interdisciplinary Research [CDO2016], the Danish National Research Foundation [DNRF107] and Wellcome Trust [098051].

Conflict of interest statement. E.P.B. declares that a patent application covering parts of the IDAA method has been filed.

REFERENCES

1. Flick KE, Jurica MS, Monnat RJ, Stoddard BL. DNA binding and cleavage by the nuclear intron-encoded homing endonuclease I-PpoI. *Nature.* 1998;394(6688):96–101.
2. Kim YG, Cha J, Chandrasegaran S. Hybrid restriction enzymes: zinc finger fusions to Fok I cleavage domain. *Proc Natl Acad Sci USA.* 1996;93(3):1156–1160.
3. Moscou MJ, Bogdanove AJ. A simple cipher governs DNA recognition by TAL effectors. *Science.* 2009;326(5959):1501.
4. Mali P, Yang L, Esvelt KM, et al. RNA-guided human genome engineering via Cas9. *Science.* 2013;339(6121):823–826.
5. Cong L, Ran FA, Cox D, et al. Multiplex genome engineering using CRISPR/Cas systems. *Science.* 2013;339(6121):819–823.
6. Symington LS, Gautier J. Double-strand break end resection and repair pathway choice. *Annu Rev Genet.* 2011;45:247–273.
7. Deriano L, Roth DB. Modernizing the nonhomologous end-joining repertoire: alternative and classical NHEJ share the stage. *Annu Rev Genet.* 2013;47:433–455.
8. Smith Gerald R. How homologous recombination is initiated: unexpected evidence for single-strand nicks from v(d)j site-specific recombination. *Cell.* 2004;117(2):146–148.
9. Stark JM, Pierce AJ, Oh J, Pastink A, Jasin M. Genetic steps of mammalian homologous repair with distinct mutagenic consequences. *Mol Cell Biol.* 2004;24(21):9305–9316.
10. Boboila C, Yan C, Wesemann DR, et al. Alternative end-joining catalyzes class switch recombination in the absence of both Ku70 and DNA ligase 4. *J Exp Med.* 2010;207 (2):417–427.
11. McVey M, Lee SE. MMEJ repair of double-strand breaks (director's cut): deleted sequences and alternative endings. *Trends Genet.* 2008;24(11):529–538.
12. Truong LN, Li Y, Shi LZ, et al. Microhomology-mediated end joining and homologous recombination share the initial end resection step to repair DNA double-strand breaks in mammalian cells. *Proc Natl Acad Sci USA.* 2013;110(19):7720–7725.
13. Moynahan ME, Jasin M. Mitotic homologous recombination maintains genomic stability and suppresses tumorigenesis. *Nat Rev Mol Cell Biol.* 2010;11(3):196–207.
14. Koike-Yusa H, Li Y, Tan E-P, Velasco-Herrera MDC, Yusa K. Genome-wide recessive genetic screening in mammalian cells with a lentiviral CRISPR-guide RNA library. *Nat Biotechnol.* 2013;32(3):267–273.
15. Yang Z, Steentoft C, Hauge C, et al. Fast and sensitive detection of indels induced by precise gene targeting. *Nucleic Acids Res.* 2015;43:e59.
16. van Overbeek M, Capurso D, Carter MM, et al. DNA repair profiling reveals nonrandom outcomes at Cas9-mediated breaks. *Mol Cell.* 2016;63(4):633–646.
17. Paquet D, Kwart D, Chen A, et al. Efficient introduction of specific homozygous and heterozygous mutations using CRISPR/Cas9. *Nature.* 2016;533(7601):125–129.
18. Lonowski LA, Narimatsu Y, Riaz A, et al. Genome editing using FACS enrichment of nuclease-expressing cells and indel detection by amplicon analysis. *Nat Protoc.* 2017;12 (3):581–603.
19. Bennett EP, Jacobi AM, Garrett RR, Behlke MA. Detection of insertion/deletion (indel) events after genome targeting: Pro's and con's of the available methods. In: Appasani K, ed. *Genome editing and Engineering: From Talens, ZFNs and CRISPRs to Molecular Surgery. Cambridge, United Kingdom: Cambridge University Press; 2018 [in press].*

20. Tan EP, Li Y, Del Castillo Velasco-Herrera M, Yusa K, Bradley A. Off-target assessment of CRISPR-Cas9 guiding RNAs in human iPS and mouse ES cells. *Genesis.* 2015;53 (2):225–236.
21. Yeung AT, Hattangadi D, Blakesley L, Nicolas E. Enzymatic mutation detection technologies. *Biotechniques.* 2005;38(5):749–758.
22. Brinkman EK, Chen T, Amendola M, van Steensel B. Easy quantitative assessment of genome editing by sequence trace decomposition. *Nucleic Acids Res.* 2014;42(22):e168.
23. Huang MC, Cheong WC, Lim LS, Li M-H. A simple, high sensitivity mutation screening using Ampligase mediated T7 endonuclease I and Surveyor nuclease with microfluidic capillary electrophoresis. *Electrophoresis.* 2012;33(5):788–796.
24. Güell M, Yang L, Church G. Genome editing assessment using CRISPR genome analyzer. *Bioinformatics.* 2014;30(20):2968–2970.
25. Pinello L, Canver MC, Hoban MD, et al. Analyzing CRISPR genome-editing experiments with CRISPResso. *Nat Biotechnol.* 2016;34(7):695–697.
26. Steentoft C, Vakhrushev SY, Joshi HJ, et al. Precision mapping of the human O-GalNAc glycoproteome through simplecell technology. *EMBO J.* 2013;32:1478–1488.
27. Hansen L, Lind-Thomsen A, Joshi HJ, et al. A glycogene mutation map for discovery of diseases of glycosylation. *Glycobiology.* 2015;25:221–224.
28. Metzakopian E, Strong A, Iyer V, Hodgkins A, et al. Enhancing the genome editing toolbox: genome wide CRISPR arrayed libraries. *Sci Rep.* 2016;7:2244.
29. Hsu PD, Scott Da, Weinstein JA, et al. DNA targeting specificity of RNA-guided Cas9 nucleases. *Nat Biotechnol.* 2013;31(9):827–832.
30. Yusa K. piggyBac transposon. *Microbiol Spectr.* 2015;3(2):1–26.
31. Mladenov E, Iliakis G. Induction and repair of DNA double strand breaks: the increasing spectrum of non-homologous end joining pathways. *Mutat Res—Fundam Mol Mech Mutagen.* 2011;711(1–2):61–72.
32. Jiang F, Doudna JA. CRISPR—Cas9 structures and mechanisms. *Annu Rev Biophys.* 2017;22(46):505–529.

CHAPTER FOUR

CRISPR Libraries and Screening

John T. Poirier[1]
Department of Medicine, Memorial Sloan Kettering Cancer Center, New York, NY, United States
[1] Corresponding author. E-mail address: poirierj@mskcc.org

Contents

Abstract

CRISPR-Cas9 technology has revolutionized large-scale functional genomic screening in mammalian cell-culture systems. Due in part to optimized lentiviral delivery vectors; it is now possible to perform CRISPR-Cas9 screens in animals in order to study biological processes in the context of a whole organism and within more physiologically relevant environment. This chapter focuses primarily on mouse models of human cancers; viral vectors used for simultaneous tumor initiation and genome editing and sgRNA library design considerations. Experience with direct and indirect in vivo RNAi screens in the literature is also discussed in order to highlight the challenges of delivering diverse libraries of small RNAs in vivo.

1. INTRODUCTION

The field of functional genomic seeks to understand the relationship between genotype and phenotype in living systems. Traditionally, candidate-gene approaches have been used to study gene function in model systems. Over the past several decades, unbiased approaches to interrogate or

Progress in Molecular Biology and Translational Science, Volume 152
ISSN 1877-1173
http://dx.doi.org/10.1016/bs.pmbts.2017.10.002

experimentally manipulate DNA sequences or gene expression levels on a genome-wide scale have opened a new frontier for discovery.

Forward genetic screens in cell lines and in rodents have historically been performed using stochastic mutagenesis techniques.[1,2] This approach has led to numerous important discoveries; however, a shortcoming of this approach is the difficulty of identifying the specific genetic alterations responsible for a given phenotype of interest due to the inability to precisely control which genes become mutagenized. A related limitation is the difficulty of mutagenizing both alleles of genes in a diploid genome; this restricted early genome-wide screening efforts in mammalian cell lines to a small number of experimental near-haploid cancer cell lines[3–5] or Bloom syndrome protein-deficient cells, which lose heterozygosity due to mitotic recombination.[6,7] In contrast, forward genetic screens that can knock down or knock out specific genes using RNA interference (RNAi) or programmable nucleases have transformed forward genetic screening in mammalian cell lines. Knock out screens have been made possible by the confluence of two disruptive technologies: CRISPR genome engineering using an RNA-programmable Cas9 nuclease and next-generation sequencing (NGS).[8,9] The former is required for gene disruption, repression, or activation while the latter enables the sequencing and relative quantitation of diverse nucleic acid libraries. Together, they constitute a powerful molecular toolbox for functional genomics in mammalian cells.

Immortalized human cell lines have played a key role in research since HeLa cells were successfully cultured in 1951.[10] Cell lines are useful for the relative ease with which they can be grown, cryopreserved, and revived, as well as the recapitulation of many of the characteristics of the primary tumors from which they are derived. Cell lines have been used in small numbers and later at industrial scales to identify relationships between genotype, gene expression, and drug sensitivity.[11–13] The advent of RNAi technology and subsequently CRISPR has demonstrated the utility of cell lines for studying cancer biology, identifying cancer essential genes, and understanding the consequences of pharmacologic perturbations.[14–16]

Cell lines grown in vitro experience a very different physiological environment and combination of selective pressures than cells proliferating in a living host. Typically, cultured cells are grown in plastic vessels containing a pH buffered, nutrient rich, liquid medium with 10% fetal bovine serum and incubated at 37°C and 5% CO_2. While this environment is ideal for the growth of most immortalized cell lines, it differs from the in vivo environment with respect to oxygen tension,[17] growth substrate,[18,19] nutrient, waste, and growth factor gradients, pH,[20,21] and the presence of stromal cells.[22] Moreover, the in vitro culture environment poses significant challenges in studying immune evasion, invasion and metastasis, inflammation, cellular energetics, tumor initiation, and maintenance, and angiogenesis. In order to study these aspects of biology, high-fidelity in vivo models are required.

In this chapter, the unique challenges of CRISPR screening—including CRISPR knockout, activation, and repression—in animals and animal models will be elaborated, focusing on mouse models of human cancers.

2. MOUSE MODELS OF CANCER

In vivo models of cancer can be divided into autochthonous models and graft models (reviewed in Ref. [22a]). Autochthonous models are mouse models in which oncogenic-somatic alterations occur spontaneously in aged animals, or are introduced either through exposure to carcinogens, or through conditional genetically engineered mouse models (GEMM).[23] These models are versatile, but may be most useful for studying early events in tumor initiation and progression, immune evasion, and the functional role of spontaneous genetic lesions observed in human cancers. CRISPR technology can be used in GEMMs in a variety of ways: introduction of Cas9 and sgRNA through viral vectors,[24–26] editing of mouse embryonic stem cells to express Cas9 and sgRNA(s),[27] liver targeting through hydrodynamic injection,[28,29] or crossing a GEMM of interest into a Cas9 expressing line.[30] GEMMs that can be conditionally initiated with lenti-Cre are perhaps the most useful because it is (1) straightforward to express sgRNAs from the same vector as Cre recombinase; (2) expressing Cas9 in the germline offers biosafety, titer, and immunological advantages; and (3) genomic integration of the lentiviral genome allows for lineage tracking.[24–26]

Immortalized mouse cancer cell lines can also be derived from GEMMs and engrafted into a syngeneic host. Such models have been referred to as GEMM-derived allografts (GDA).[23] Human cancer cell lines play a role here, as well as they can be xenografted into a variety of different immune compromised hosts to form tumors. Whether human or mouse, cell lines can be engrafted into hosts through different routes and anatomical contexts to mimic systemic dissemination, orthotopic growth in the organ of origin, or most commonly growth of a subcutaneous mass.

3. CRISPR-CAS9 SYSTEMS

CRISPR-Cas9 genome editing depends on the nuclear localization of a ribonucleoprotein (RNP) complex comprising Cas9 and a single guide RNA (sgRNA). Cas9:sgRNA RNPs can be delivered directly through a variety of approaches or indirectly by introducing nucleic acid sequences encoding Cas9 and an sgRNA of interest.[31–34] Transient introduction of RNA or RNPs can provide higher editing specificity than other approaches due to the short time window in which they are expressed. A variety of viral vectors have been developed for stable expression of Cas9 and sgRNAs from a single vector; however, this approach is complicated by low-achievable viral titer as effective titer of lentiviral vectors decreases with increased genome size.[35] The most widely used approaches now employ sequential infection with independent lentiviruses encoding Cas9, which constitutes the major burden on packaging limit, followed by sgRNA of interest which are smaller and can be produced at higher titer.[36] The primary effects of high-titer production of the sgRNA component are the increased ease of library production and maintenance of library diversity.

While it is straightforward to transduce cells with Cas9 vectors in vitro and obtain a population of cells expressing Cas9 through drug selection, this approach is not tractable in vivo. Germline integration of Cas9 has been used as an alternative approach.[27,30] Cre-dependent Cas9 knockin mice were generated by inserting Cas9 into the Rosa26 locus.[30] Conditional control is achieved by inserting a strong 3× polyA signal upstream of Cas9 flanked by loxP sites, commonly referred to as a lox-stop-lox (LSL) approach. Cas9 expression is readily induced by the expression of Cre recombinase, which excises the 3× polyA signal. In an alternative approach Cas9 was inserted into the Col1a1 locus under the control of a doxycycline inducible promoter using FLP-FRT recombination.[27] The latter approach is unique from the former in two respects: (1) while LSL-Cas9 expression is irreversible once activated by Cre, doxycycline inducible systems are fully reversible by withdrawal of docycycline; and (2) it allows simultaneous integration of one or more sgRNAs under the control of a constitutive promoter. This could potentially allow for finer tuned control of genome editing by precisely defining the temporal window in which Cas9 is expressed.

With precise control of genome editing in mind, a number of variants have been developed that can be induced by doxycycline,[27,37–40] tamoxifen,[41,42] rapamycin,[43] and light.[44,45] Each of these approaches is intended to provide fine control of Cas9 expression. The ability to restrict genome editing to a specific temporal window is essential for studying many processes in vivo, such as the execution of developmental programs, precisely timed tumor initiation, or the identification of genes required for maintenance of already established tumors. It should be noted that each of the inducible systems described previously, is subject to particular strengths and weaknesses with respect to leakiness, switching kinetics, and magnitude of induction.

In addition to gene disruption by Cas9 nuclease, several variants based on fusions of a catalytically dead Cas9 mutant (dCas9) to transcriptional activators (CRISPRa) or inactivators (CRISPRi) have been developed.[46–50] Among CRISPRa approaches, the synergistic activation mediator (SAM) system, which recruits p65 and HSF1 activators to a dCas9-VP64 fusion via MS2-binding stem loops engineered into the sgRNA sequence, has demonstrated the greatest degree of activation.[49,51] Genome-scale CRISPRa screens have therefore primarily employed this system.[49] Relatively few in vivo screens have been attempted with CRISPRi and CRISPRa compared to CRISPR knockout; however, CRISPR activation and repression screens have been reported in indirect in vivo screens using syngeneic mouse models.[52]

4. CRISPR LIBRARIES

For the purposes of this chapter, a screen will constitute, an experiment in which, multiple sgRNAs, a library, are tested and ranked by their biological effect. In general, libraries may come in two formats: arrayed format, in which each element of the library is compartmentalized, and pooled format, in which each element of the library is present in a single pool. Arrayed libraries can be advantageous for studying the effect of gene disruption on complex phenotypes using; for example, high-content imaging systems. Arrayed experiments can also be used to study subtle phenotypes that require parallel analyses of thousands of cells each of which has the same perturbation, such as a cell cycle dependent phenotype. Conversely, pooled formats are relatively inexpensive to perform, and require little specialized equipment, but are typically used for screens in which the readout is an acquired fitness advantage. This chapter will focus on the use of pooled libraries.

Pooled screens are dependent on the controlled distribution of sgRNAs to a population of cells such that reach infected cell incorporates a single sgRNA in order to prevent the confounding effects of multiply transduced cells. This is achieved by carefully controlling the multiplicity of infection (MOI): the ratio of infectious lentivirus to target cells. During an infection, viruses are distributed to cells based on Poisson loading; thus at a low MOI of 0.3–0.5 transduction units/cell, the majority of infected cells are transduced with at most one sgRNA.

The frequency of guides in a high-quality pooled sgRNA library follows an approximately normal distribution. Ideally, under- or overrepresented guides should be no more than one log above or below the mean of the distribution. Underrepresented guides may be missed in a screen due to their overall scarcity, while irrelevant overrepresented guides may dominate a screen by virtue of their initial abundance.

Pooled screens exploit an experimentally controlled selective pressure to alter the representation of sgRNAs with relevance to a specific phenotype.[53] Positive selection experiments are designed to enrich for sgRNAs that increase fitness, while negative selection experiments seek to detect sgRNAs that drop out of the population due to loss of fitness.[54] In a positive selection screen, the diversity of the library is greatly reduced due to elimination of cells harboring sgRNAs that are irrelevant to the phenotype. The distribution of sgRNAs can be expected to skew drastically to overrepresentation of a relatively small number of guides, typically exclusive of nontargeting, negative control sgRNAs. Positive selection experiments are by their nature more robust due to the nature of high signal to noise ratio.[55] Conversely negative selection screens eliminate sgRNAs that affect the phenotype of interest. This type of screen will typically have much higher sgRNA diversity at the end of the experiment.

Pooled CRISPR libraries can be obtained directly from investigators, nonprofit organizations, and commercial sources in many formats ranging from bacterial plasmid libraries to ready-to-transduce lentiviral pools.

4.1 Library Scale and Diversity

The scale of an sgRNA library can span multiple orders of magnitude: published libraries range from 10^1 to 10^5 unique sgRNAs. Total library diversity can exceed this range when molecular barcodes are employed, as describe later in this chapter. At the higher end of the diversity range are genome-scale libraries targeting every gene in the human or mouse genome. These libraries are of appropriate complexity for screening in vitro; however, they contain too many elements for practical screening in vivo. In vivo screens typically rely on smaller, focused libraries or sublibraries. In one large-scale screen, a cell pool was mutagenized in vitro, then injected into an immunocompromised host to select for clones with higher metastatic potential.[56] Interestingly, only sgRNAs targeting a small number of genes were recovered. This screen highlighted the stringency of screens of this nature, as well as the challenge of identifying subtle phenotypes from among the more pronounced.

It is instructive to revisit the RNAi literature when highlighting the advantages and limitations of screening diverse libraries of small RNAs in vivo. Numerous in vivo screens have been reported, often with unexpected results.[57–72] A common theme among these papers was the surprising number of shRNAs that could not be recovered at the conclusion of the experiment; this has been attributed to either stochastic clonal outgrowth or a surprisingly high fraction of essential genes unique to the in vivo setting.

4.2 sgRNA Design

First generation sgRNA libraries were designed before robust rules for predicting sgRNA activity had been developed. Selective and potent sgRNAs are required to minimize signal from off-target effects and ensure that independent guides targeting the same gene produce consistent phenotypes. It follows that low-activity sgRNAs can dilute the biological signal in gene-wise analyses. This becomes particularly important when screening libraries with sparse sgRNA coverage per gene. Iterative refinement of these rules has increased on-target sgRNA activity while simultaneously reducing off-target effects, resulting in improved second- and third-generation libraries.[73,74]

It is equally important to consider these factors when selecting sgRNA sequences for a custom library or validation experiment. Ideally, a given sgRNA should cut a unique sequence in the genome of interest, or minimize the number of potential off-target sites. The majority of sgRNA design tools rely on short read aligners to enumerate the number of nonunique sites for a given number of mismatches. Unfortunately, short read aligners are not exhaustive in this respect and may report a given guide as unique even though it cuts at multiple loci. GuideScan is a tool that comprehensively annotates potential off target effects while incorporating the high-performing activity scoring algorithm described previously.[75]

The precise location within the structure of a gene is also important. In the context of CRISPR-Cas9 knockout screens, which exert their function when out of frame insertions and deletions lead to early stop codons, sgRNAs should be located within the 5′ most constitutively expressed coding exon.[8] In contrast, CRISPR activation or repression libraries should focus guides near the transcriptions tart site (TSS). Additional considerations include avoidance of homopolymer sequences or sequences with strong GC bias. If there is prior knowledge of the type of functional domain that a researcher may wish to target, it has also been shown to be highly efficient to target these domains directly.[76]

One final consideration for sgRNA design is biosafety. When using viral vectors, separation of Cas9 from the targeting sgRNA is preferred to reduce the risk of genome editing in laboratory personnel. This is especially critical when sgRNAs are designed to target tumor suppressor genes or activate oncogenes. Wherever possible sgRNA designed to target genes in mice should be selected not to target human genomic sequences.

4.3 Molecular Barcoding

In the simplest case, a researcher may wish to measure only the changes in abundance of a given sgRNA to study gene function. However, when studying the effects of a specific sgRNA on tumor initiation and progression in vivo, the abundance of an sgRNA sequence can be a function of the initial representation in the library, the number of initiation events, and the overall tumor burden, which is itself influenced by a variety of factors including proliferation rate, angiogenic potential, and resistance to apoptosis. These factors influencing abundance can be controlled for using unique molecular barcodes to track clonal dynamics.[77–79]

Molecular barcodes are random nucleotides sequence introduced into a library, typically by amplification with PCR primers containing some number of degenerate bases. Each individual element in the library is therefore associated with a diverse number of different barcode sequences. This approach can increase the total library diversity by orders of magnitude and is therefore only possible when the number of elements of interest in the library is relatively small. Importantly, since molecular barcodes are incorporated into the genome, they are subject to error introduced through normal mutagenic processes, as well as sequencing error rates. It is critical to distinguish between barcodes that are truly representing distinct clones and those that arise from sources of sequence drift.[80]

A strategy in which sgRNA are paired to molecular barcodes allows for unambiguous identification of individual tumor initiating events. This approach has been applied to an autochthonous lung cancer model to allow measurement of both the number of lesions and tumor burden in a given animal, but could be equally applicable to any GEMM in which a lentiviral vector is used for tumor initiation.[81]

Graft models are also subject to clonal dynamics during engraftment. As described previously, engrafted tumors are often generated by a minority of seeded clones; it can therefore be impossible to distinguish between enrichment of a given sgRNA due to clonal jackpotting as opposed to increased fitness, especially in the context of a constitutively expressed Cas9. This effect can be partially controlled for by assessing enrichment of multiple independent sgRNAs targeting the same gene; however, this method is susceptible to the confounding effects of sgRNAs with low activity. Molecular barcoding in this case allows for simultaneous assessment of clonal dynamics and fitness.

4.4 Essential Controls

Control sgRNAs are essential to any screening experiment. At a minimum, each screen should be performed with abundant negative control sgRNAs that are expected to have no biological impact. Published libraries have used as control sgRNAs with no homology to the target genome. Such nontargeting controls are useful; however, the effect of these sgRNAs is not equivalent to targeting sgRNAs since they don't induce double strand breaks. This is especially important when guides target multiple areas of the genome due to copy number amplification or high off-target activity.[82–85] A complementary approach is to employs sgRNAs that target intragenic regions of the genome: so-called "safe-targeting" sgRNAs.[86]

It is also helpful to have positive-control sgRNAs. These sgRNAs can take the form of guides targeting core essential genes in the case of a drop out screen or known genes of the same class. For example, in a candidate tumor suppressor screen it would be useful to include classic tumor suppressor genes, such as *TP53, RB1*, and *PTEN* as controls.

4.5 DNA Extraction and Library Amplification

DNA extraction and PCR amplification steps are often overlooked aspects of CRISPR screens since they are common laboratory techniques. However, these steps are performed at a larger scale than what is typically performed by the average researcher, presenting unique challenges. Since many aspects of this step cannot be readily assessed for quality control until after library sequencing, it can be time consuming and wasteful to optimize. Therefore, using published protocols is highly recommended.[53]

Extraction of clean, high-quality genomic DNA from large amounts of tumor tissue is essential. Tumor DNA used for PCR amplification must be free of PCR inhibitors or contaminants that could prevent accurate determination of DNA concentration. The approach used must also be compatible with processing large amounts of starting material, which favors the application of a precipitation-based technique, but it should also be noted that carryover must be minimized.

Attention should be paid to minimizing burden DNA in the sample prior to amplification. Burden DNA is the DNA mass that is derived from tissues other than the cancer cells of interest. In the case of flank xenografts, irrelevant burden DNA of murine stromal origin can make up more than half of the total nucleic acids.[87] Efforts should be made to remove this infiltrating stroma.[88] In GEMMs, tumor cells—often tens or hundreds of distinct lesions—are admixed with normal organ parenchyma. Marking these cells with a fluorescent marker for fluorescence activated cell sorting (FACS) is recommended to obtain a pure tumor cell population.[89]

5. CONCLUDING REMARKS

Many aspects of cancer biology can only be modeled with high fidelity in animals. Animal models of cancer will therefore remain as important research platforms in the coming years. Despite being in its relative infancy, CRISPR-Cas9 genome editing technology is fast becoming an essential tool for studying functional genomics across many fields. Coupled with the power of next generation sequencing to interrogate highly diverse pools of nucleic acids, CRISPR-Cas9 screening is emerging as a transformative tool for unraveling cancer biology. The application of CRISPR-Cas9 for discovery screens will continue to expand our understanding of these diverse processes in vivo.

REFERENCES

1. Grimm S. The art and design of genetic screens: mammalian culture cells. *Nat Rev Genet.* 2004;5(3):179–189.
2. Kile BT, Hilton DJ. The art and design of genetic screens: mouse. *Nat Rev Genet.* 2005;6(7):557–567.
3. Blomen VA, Majek P, Jae LT, et al. Gene essentiality and synthetic lethality in haploid human cells. *Science.* 2015;350(6264):1092–1096.
4. Jae LT, Raaben M, Riemersma M, et al. Deciphering the glycosylome of dystroglycanopathies using haploid screens for lassa virus entry. *Science.* 2013;340(6131):479–483.

5. Carette JE, Guimaraes CP, Varadarajan M, et al. Haploid genetic screens in human cells identify host factors used by pathogens. *Science.* 2009;326(5957):1231–1235.
6. Traverso G, Bettegowda C, Kraus J, et al. Hyper-recombination and genetic instability in BLM-deficient epithelial cells. *Cancer Res.* 2003;63(24):8578–8581.
7. Guo G, Wang W, Bradley A. Mismatch repair genes identified using genetic screens in Blm-deficient embryonic stem cells. *Nature.* 2004;429(6994):891–895.
8. Shalem O, Sanjana NE, Hartenian E, et al. Genome-scale CRISPR-Cas9 knockout screening in human cells. *Science.* 2014;343(6166):84–87.
9. Shalem O, Sanjana NE, Zhang F. High-throughput functional genomics using CRISPR-Cas9. *Nat Rev Genet.* 2015;16(5):299–311.
10. Masters JR. HeLa cells 50 years on: the good, the bad and the ugly. *Nat Rev Cancer.* 2002;2(4):315–319.
11. Alley MC, Scudiero DA, Monks A, et al. Feasibility of drug screening with panels of human tumor cell lines using a microculture tetrazolium assay. *Cancer Res.* 1988;48(3):589–601.
12. Park ES, Rabinovsky R, Carey M, et al. Integrative analysis of proteomic signatures, mutations, and drug responsiveness in the NCI 60 cancer cell line set. *Mol Cancer Ther.* 2010;9(2):257–267.
13. Barretina J, Caponigro G, Stransky N, et al. The Cancer Cell Line Encyclopedia enables predictive modelling of anticancer drug sensitivity. *Nature.* 2012;483(7391):603–607.
14. Hannon GJ, Conklin DS. RNA interference by short hairpin RNAs expressed in vertebrate cells. *Methods Mol Biol.* 2004;257:255–266.
15. Hannon GJ. RNA interference. *Nature.* 2002;418(6894):244–251.
16. Sander JD, Joung JK. CRISPR-Cas systems for editing, regulating and targeting genomes. *Nat Biotechnol.* 2014;32(4):347–355.
17. Carreau A, El Hafny-Rahbi B, Matejuk A, Grillon C, Kieda C. Why is the partial oxygen pressure of human tissues a crucial parameter? Small molecules and hypoxia. *J Cell Mol Med.* 2011;15(6):1239–1253.
18. Allen LT, Tosetto M, Miller IS, et al. Surface-induced changes in protein adsorption and implications for cellular phenotypic responses to surface interaction. *Biomaterials.* 2006;27(16):3096–3108.
19. Gilkes DM, Semenza GL, Wirtz D. Hypoxia and the extracellular matrix: drivers of tumour metastasis. *Nat Rev Cancer.* 2014;14(6):430–439.
20. Hirschhaeuser F, Sattler UG, Mueller-Klieser W. Lactate: a metabolic key player in cancer. *Cancer Res.* 2011;71(22):6921–6925.
21. Parks SK, Chiche J, Pouyssegur J. Disrupting proton dynamics and energy metabolism for cancer therapy. *Nat Rev Cancer.* 2013;13(9):611–623.
22. Bhowmick NA, Neilson EG, Moses HL. Stromal fibroblasts in cancer initiation and progression. *Nature.* 2004;432(7015):332–337.

22a. Day CP, Merlino G, Van Dyke T. Preclinical mouse cancer models: a maze of opportunities and challenges. *Cell.* 2015;163(1):39–53.

23. Heyer J, Kwong LN, Lowe SW, Chin L. Non-germline genetically engineered mouse models for translational cancer research. *Nat Rev Cancer.* 2010;10(7):470–480.
24. Sanchez-Rivera FJ, Papagiannakopoulos T, Romero R, et al. Rapid modelling of cooperating genetic events in cancer through somatic genome editing. *Nature.* 2014;516(7531):428–431.
25. Chiou SH, Winters IP, Wang J, et al. Pancreatic cancer modeling using retrograde viral vector delivery and in vivo CRISPR/Cas9-mediated somatic genome editing. *Genes Dev.* 2015;29(14):1576–1585.
26. Annunziato S, Kas SM, Nethe M, et al. Modeling invasive lobular breast carcinoma by CRISPR/Cas9-mediated somatic genome editing of the mammary gland. *Genes Dev.* 2016;30(12):1470–1480.

27. Dow LE, Fisher J, O'Rourke KP, et al. Inducible in vivo genome editing with CRISPR-Cas9. *Nat Biotechnol.* 2015;33(4):390–394.
28. Yin H, Xue W, Chen S, et al. Genome editing with Cas9 in adult mice corrects a disease mutation and phenotype. *Nat Biotechnol.* 2014;32(6):551–553.
29. Xue W, Chen S, Yin H, et al. CRISPR-mediated direct mutation of cancer genes in the mouse liver. *Nature.* 2014;514(7522):380–384.
30. Platt RJ, Chen S, Zhou Y, et al. CRISPR-Cas9 knockin mice for genome editing and cancer modeling. *Cell.* 2014;159(2):440–455.
31. Liu J, Gaj T, Yang Y, et al. Efficient delivery of nuclease proteins for genome editing in human stem cells and primary cells. *Nat Protoc.* 2015;10(11):1842–1859.
32. Cong L, Ran FA, Cox D, et al. Multiplex genome engineering using CRISPR/Cas systems. *Science.* 2013;339(6121):819–823.
33. Mali P, Yang L, Esvelt KM, et al. RNA-guided human genome engineering via Cas9. *Science.* 2013;339(6121):823–826.
34. Wang H, Yang H, Shivalila CS, et al. One-step generation of mice carrying mutations in multiple genes by CRISPR/Cas-mediated genome engineering. *Cell.* 2013;153(4): 910–918.
35. Kumar M, Keller B, Makalou N, Sutton RE. Systematic determination of the packaging limit of lentiviral vectors. *Hum Gene Ther.* 2001;12(15):1893–1905.
36. Sanjana NE, Wright J, Zheng K, et al. High-resolution interrogation of functional elements in the noncoding genome. *Science.* 2016;353(6307):1545–1549.
37. Gonzalez F, Zhu Z, Shi ZD, et al. An iCRISPR platform for rapid, multiplexable, and inducible genome editing in human pluripotent stem cells. *Cell Stem Cell.* 2014;15(2): 215–226.
38. Zhu Z, Gonzalez F, Huangfu D. The iCRISPR platform for rapid genome editing in human pluripotent stem cells. *Methods Enzymol.* 2014;546:215–250.
39. Chen Y, Cao J, Xiong M, et al. Engineering human stem cell lines with inducible gene knockout using CRISPR/Cas9. *Cell Stem Cell.* 2015;17(2):233–244.
40. Aubrey BJ, Kelly GL, Kueh AJ, et al. An inducible lentiviral guide RNA platform enables the identification of tumor-essential genes and tumor-promoting mutations in vivo. *Cell Rep.* 2015;10(8):1422–1432.
41. Davis KM, Pattanayak V, Thompson DB, Zuris JA, Liu DR. Small molecule-triggered Cas9 protein with improved genome-editing specificity. *Nat Chem Biol.* 2015;11(5): 316–318.
42. Liu KI, Ramli MN, Woo CW, et al. A chemical-inducible CRISPR-Cas9 system for rapid control of genome editing. *Nat Chem Biol.* 2016;12(11):980–987.
43. Zetsche B, Volz SE, Zhang F. A split-Cas9 architecture for inducible genome editing and transcription modulation. *Nat Biotechnol.* 2015;33(2):139–142.
44. Polstein LR, Gersbach CA. A light-inducible CRISPR-Cas9 system for control of endogenous gene activation. *Nat Chem Biol.* 2015;11(3):198–200.
45. Nihongaki Y, Yamamoto S, Kawano F, Suzuki H, Sato M. CRISPR-Cas9-based photoactivatable transcription system. *Chem Biol.* 2015;22(2):169–174.
46. Maeder ML, Linder SJ, Cascio VM, Fu Y, Ho QH, Joung JK. CRISPR RNA-guided activation of endogenous human genes. *Nat Methods.* 2013;10(10):977–979.
47. Qi LS, Larson MH, Gilbert LA, et al. Repurposing CRISPR as an RNA-guided platform for sequence-specific control of gene expression. *Cell.* 2013;152(5):1173–1183.
48. Gilbert LA, Horlbeck MA, Adamson B, et al. Genome-scale CRISPR-mediated control of gene repression and activation. *Cell.* 2014;159(3):647–661.
49. Konermann S, Brigham MD, Trevino AE, et al. Genome-scale transcriptional activation by an engineered CRISPR-Cas9 complex. *Nature.* 2015;517(7536):583–588.
50. Perez-Pinera P, Kocak DD, Vockley CM, et al. RNA-guided gene activation by CRISPR-Cas9-based transcription factors. *Nat Methods.* 2013;10(10):973–976.

51. Chavez A, Tuttle M, Pruitt BW, et al. Comparison of Cas9 activators in multiple species. *Nat Methods.* 2016;13(7):563–567.
52. Braun CJ, Bruno PM, Horlbeck MA, Gilbert LA, Weissman JS, Hemann MT. Versatile in vivo regulation of tumor phenotypes by dCas9-mediated transcriptional perturbation. *Proc Natl Acad Sci USA.* 2016;113(27):E3892–E3900.
53. Joung J, Konermann S, Gootenberg JS, et al. Genome-scale CRISPR-Cas9 knockout and transcriptional activation screening. *Nat Protoc.* 2017;12(4):828–863.
54. Miles LA, Garippa RJ, Poirier JT. Design execution, and analysis of pooled in vitro CRISPR/Cas9 screens. *FEBS J.* 2016;283(17):3170–3180.
55. Kaelin Jr WG. Common pitfalls in preclinical cancer target validation. *Nat Rev Cancer.* 2017;17(7):425–440.
56. Chen S, Sanjana NE, Zheng K, et al. Genome-wide CRISPR screen in a mouse model of tumor growth and metastasis. *Cell.* 2015;160(6):1246–1260.
57. Scuoppo C, Miething C, Lindqvist L, et al. A tumour suppressor network relying on the polyamine-hypusine axis. *Nature.* 2012;487(7406):244–248.
58. Miller PG, Al-Shahrour F, Hartwell KA, et al. In Vivo RNAi screening identifies a leukemia-specific dependence on integrin beta 3 signaling. *Cancer Cell.* 2013;24(1): 45–58.
59. Beronja S, Janki P, Heller E, et al. RNAi screens in mice identify physiological regulators of oncogenic growth. *Nature.* 2013;501(7466):185–190.
60. Wuestefeld T, Pesic M, Rudalska R, et al. A Direct in vivo RNAi screen identifies MKK4 as a key regulator of liver regeneration. *Cell.* 2013;153(2):389–401.
61. Possik PA, Muller J, Gerlach C, et al. Parallel in vivo and in vitro melanoma RNAi dropout screens reveal synthetic lethality between hypoxia and DNA damage response inhibition. *Cell Rep.* 2014;9(4):1375–1386.
62. Chen R, Belanger S, Frederick MA, et al. In vivo RNA interference screens identify regulators of antiviral CD4(+) and CD8(+) T cell differentiation. *Immunity.* 2014;41 (2):325–338.
63. Schramek D, Sendoel A, Segal JP, et al. Direct in vivo RNAi screen unveils myosin IIa as a tumor suppressor of squamous cell carcinomas. *Science.* 2014;343(6168):309–313.
64. Meacham CE, Lawton LN, Soto-Feliciano YM, et al. A genome-scale in vivo loss-of-function screen identifies Phf6 as a lineage-specific regulator of leukemia cell growth. *Genes Dev.* 2015;29(5):483–488.
65. Baratta MG, Schinzel AC, Zwang Y, et al. An in-tumor genetic screen reveals that the BET bromodomain protein, BRD4, is a potential therapeutic target in ovarian carcinoma. *Proc Natl Acad Sci USA.* 2015;112(1):232–237.
66. Bossi D, Cicalese A, Dellino GI, et al. In vivo genetic screens of patient-derived tumors revealed unexpected frailty of the transformed phenotype. *Cancer Discov.* 2016;6 (6):650–663.
67. Caino MC, Seo JH, Aguinaldo A, et al. A neuronal network of mitochondrial dynamics regulates metastasis. *Nat Commun.* 2016;7:13730.
68. Zhou P, Shaffer DR, Alvarez Arias DA, et al. In vivo discovery of immunotherapy targets in the tumour microenvironment. *Nature.* 2014;506(7486):52–57.
69. Malina A, Mills JR, Cencic R, et al. Repurposing CRISPR/Cas9 for in situ functional assays. *Genes Dev.* 2013;27(23):2602–2614.
70. Zender L, Xue W, Zuber J, et al. An oncogenomics-based in vivo RNAi screen identifies tumor suppressors in liver cancer. *Cell.* 2008;135(5):852–864.
71. Rudalska R, Dauch D, Longerich T, et al. In vivo RNAi screening identifies a mechanism of sorafenib resistance in liver cancer. *Nat Med.* 2014;20(10):1138–1146.
72. Parnas O, Jovanovic M, Eisenhaure TM, et al. A genome-wide CRISPR screen in primary immune cells to dissect regulatory networks. *Cell.* 2015;162(3): 675–686.

73. Doench JG, Fusi N, Sullender M, et al. Optimized sgRNA design to maximize activity and minimize off-target effects of CRISPR-Cas9. *Nat Biotechnol.* 2016;34(2):184–191.
74. Doench JG, Hartenian E, Graham DB, et al. Rational design of highly active sgRNAs for CRISPR-Cas9-mediated gene inactivation. *Nat Biotechnol.* 2014;32(12):1262–1267.
75. Perez AR, Pritykin Y, Vidigal JA, et al. GuideScan software for improved single and paired CRISPR guide RNA design. *Nat Biotechnol.* 2017;35(4):347–349.
76. Shi J, Wang E, Milazzo JP, Wang Z, Kinney JB, Vakoc CR. Discovery of cancer drug targets by CRISPR-Cas9 screening of protein domains. *Nat Biotechnol.* 2015;33 (6):661–667.
77. Bhang HE, Ruddy DA, Krishnamurthy Radhakrishna V, et al. Studying clonal dynamics in response to cancer therapy using high-complexity barcoding. *Nat Med.* 2015;21 (5):440–448.
78. Levy SF, Blundell JR, Venkataram S, Petrov DA, Fisher DS, Sherlock G. Quantitative evolutionary dynamics using high-resolution lineage tracking. *Nature.* 2015;519 (7542):181–186.
79. Nguyen LV, Pellacani D, Lefort S, et al. Barcoding reveals complex clonal dynamics of de novo transformed human mammary cells. *Nature.* 2015;528(7581):267–271.
80. Callahan BJ, McMurdie PJ, Rosen MJ, Han AW, Johnson AJ, Holmes SP. DADA2: high-resolution sample inference from Illumina amplicon data. *Nat Methods.* 2016;13 (7):581–583.
81. Rogers ZN, McFarland CD, Winters IP, et al. A quantitative and multiplexed approach to uncover the fitness landscape of tumor suppression in vivo. *Nat Methods.* 2017;14 (7):737–742.
82. Wang T, Birsoy K, Hughes NW, et al. Identification and characterization of essential genes in the human genome. *Science.* 2015;350(6264):1096–1101.
83. Aguirre AJ, Meyers RM, Weir BA, et al. Genomic copy number dictates a gene-independent cell response to CRISPR/Cas9 targeting. *Cancer Discov.* 2016;6 (8):914–929.
84. Munoz DM, Cassiani PJ, Li L, et al. CRISPR screens provide a comprehensive assessment of cancer vulnerabilities but generate false-positive hits for highly amplified genomic regions. *Cancer Discov.* 2016;6(8):900–913.
85. Wang T, Yu H, Hughes NW, et al. Gene essentiality profiling reveals gene networks and synthetic lethal interactions with oncogenic Ras. *Cell.* 2017;168(5). 890–903 e815.
86. Morgens DW, Wainberg M, Boyle EA, et al. Genome-scale measurement of off-target activity using Cas9 toxicity in high-throughput screens. *Nat Commun.* 2017;8:15178.
87. Lin MT, Tseng LH, Kamiyama H, et al. Quantifying the relative amount of mouse and human DNA in cancer xenografts using species-specific variation in gene length. *Biotechniques.* 2010;48(3):211–218.
88. Schneeberger VE, Allaj V, Gardner EE, Poirier JT, Rudin CM. Quantitation of murine stroma and selective purification of the human tumor component of patient-derived xenografts for genomic analysis. *PLoS One.* 2016;11(9):e0160587.
89. Madisen L, Zwingman TA, Sunkin SM, et al. A robust and high-throughput Cre reporting and characterization system for the whole mouse brain. *Nat Neurosci.* 2010;13(1):133–140.

CHAPTER FIVE

Genome Engineering Using Haploid Embryonic Stem Cells

Takuro Horii[1], Izuho Hatada
Biosignal Genome Resource Center, Institute for Molecular and Cellular Regulation, Gunma University, Maebashi, Gunma, Japan
[1] Corresponding author. E-mail address: horii@gunma-u.ac.jp

Contents

Abstract

Haploidy is a useful feature for the study of gene function because disruption of one allele in haploid cells, which contain only a single set of chromosomes, can cause loss-of-function phenotypes. Recent success in generating haploid embryonic stem (ES) cells from several mammalian species, including human, provides a new platform for simple genetic manipulation of the mammalian genome. The genome-editing potential of the CRISPR/Cas system is enhanced by the use of haploid ES cells. For example, CRISPR/Cas has been used for high-efficiency generation of multiple knockouts and knockins in haploid ES cells, with potential application in genome-wide screening. In addition, live mice can be successfully obtained by nuclear transplantation of haploid ES cells into oocytes, providing a novel approach for the generation of gene-modified animals.

Progress in Molecular Biology and Translational Science, Volume 152
ISSN 1877-1173
http://dx.doi.org/10.1016/bs.pmbts.2017.09.001

1. INTRODUCTION

The term ploidy refers to the number of sets of chromosomes in a cell. Most animal cells are diploid, containing two chromosome sets. For genetic screening of drug-resistance or disease-related genes, haploid cells, which contain a single set of chromosomes, are more useful than diploid cells. Moreover, for gene function analysis, it is necessary to disrupt both copies of a gene locus in diploid cells, as heterozygous mutations frequently do not result in phenotypic change (Fig. 1). This substantially adds to the complexity of gene function analysis in diploid cells. Conventional gene knockouts are produced either by selecting targeted clones exhibiting loss-of-heterozygosity, or by sequential targeting of both chromosomes using different resistance genes.[1,2] By contrast, disruption of one allele can generate a loss-of-function phenotype in haploid cells, as they contain only a single copy of each chromosome (Fig. 1). Hence, gene manipulation in haploid cells can simplify genetic analysis and has the potential to provide valuable information regarding the molecular functions of numerous genes and proteins. However, haploid cells generally have limited life spans; and in mammals, only gametes are haploid. Therefore, considerable efforts have been required to establish haploid mammalian cell lines.

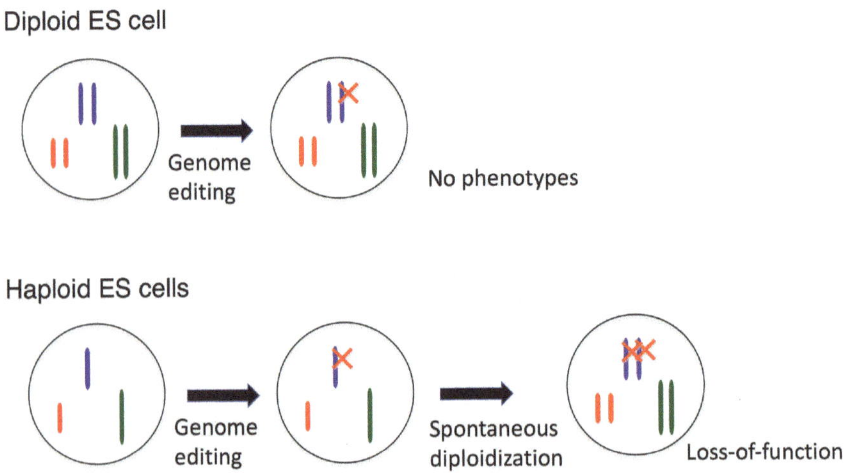

Fig. 1 Haploid embryonic stem (ES) cells are useful for studying gene function, because disruption of only one allele can cause loss-of-function phenotypes.

In 1970, the first vertebrate haploid cell lines were established from androgenetic frog embryos, generated by removing maternal nuclei from zygotes.[3] More recently, haploid embryonic stem (ES) cell lines were also produced from parthenogenetic embryos of medaka fish.[4] These reports indicate that haploid cells can be established not only from lower vertebrates, but also from organisms of similar complexity to mammals. Since 1970s, a great deal of effort has been expended in attempts to establish haploid cell lines from mouse embryos; however, for decades no proliferative haploid cell lines were obtained.[5–8] After establishment of diploid ES cells in 1981, numerous attempts to establish haploid ES cells were made in vain, because haploid cells invariably spontaneously reverted to the diploid state during ES cell establishment.[9] Finally, successful generation of mouse haploid ES cells was reported 30 years after the first report of diploid ES cells.[10,11]

2. GENERATION OF HAPLOID ES CELLS

Success in generating mouse haploid ES cells was due to two major technical modifications of the ES cell culture system: (1) culture conditions, including inhibitors of glycogen synthase kinase 3 (GSK3) and mitogen-activated protein kinase (MAPK) (collectively termed 2i); and (2) enrichment of the haploid population by fluorescence-activated cell sorting (FACS) during derivation and maintenance of ES cell cultures.

Using this system, the first mammalian haploid ES cells were generated from parthenogenetic embryos produced by artificial activation, followed by extrusion of the second polar body (Fig. 2).[10,11] Haploid ES cells were also generated from androgenetic embryos by removal of the maternal pronucleus[12] or by sperm injection into enucleated oocytes (Fig. 2).[12,13] Haploid ES cells can differentiate into functional tissues both in vitro and in chimeric mice;[10–14] however, all haploid-derived cells, differentiated tissues, and adult chimeric mice are diploid. To date, complete haploid ES cells have been generated in several additional species, including rat,[15] monkey,[16] and human.[17]

3. PURIFICATION AND MAINTENANCE OF HAPLOID CELLS

The spontaneous diploidization of haploid ES cells tends to occur by endoreduplication or omission of cytokinesis, rather than cell fusion.[18] In general, haploid ES cells are completely diploidized within 3 weeks;

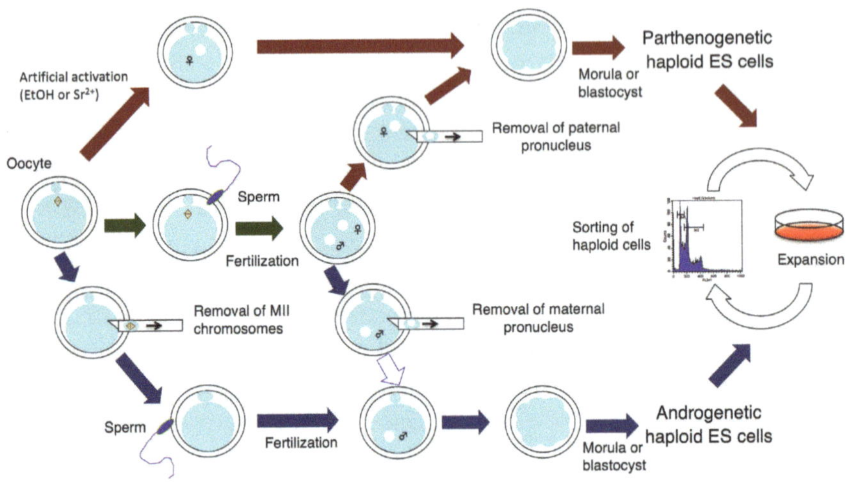

Fig. 2 Generation of haploid ES cells from parthenogenetic and androgenetic haploid embryos.

therefore, it is necessary to purify haploid cells at specific intervals (Fig. 2). Enrichment of haploid cells at G1 phase is generally performed by cell sorting after staining with Hoechst 33342; however, this method requires a FACS system with a 375 nm near-ultraviolet (UV) laser, with which ordinary FACS systems are rarely equipped. To overcome this issue, a novel haploid purification method based on cell size has recently been developed, using forward and side scatter plots to distinguish cell size and internal complexity.[19,20] During the G1 phase of the cell cycle, haploid ES cells have smaller diameters than their diploid counterparts; hence, scatter plots can be used to distinguish cells with haploid content from those with other ploidy levels at G1. Cell sorting depending on scatter plot analysis requires only a standard 488 nm laser, rather than the exceptional 375 nm lasers needed for the original method. Furthermore, Hoechst 33342 staining is not necessary for this sorting system.

It is also important to inhibit spontaneous diploidization during routine culture of haploid ES cells. The stability of the haploid karyotype is increased when ES cells are in the naïve base state; therefore, 2i, which sustain the naïve pluripotent base state, are generally used to supplement the culture medium as important inhibitors of diploidization.[18] By contrast, exit from naïve state to a primed state (e.g., FGF/activin supplementation) induces diploidization.[12] Also, during the process of ES cell differentiation, dosage compensation occurs in female diploid cells (XX) via *Xist* expression and heterochromatic inactivation of one X chromosome.[15] In these diploid cells, the ratio of X-linked (X) to autosomal (A) gene expression is 1:2. By contrast, the X:A gene expression ratio in female haploid cells is 1:1. Differentiation of haploid mouse cells lacking dosage compensation may lead to unstable cell survival and stimulate diploidization. Therefore, supplementation of growth medium with 2i is key to inhibition of diploidization in nonhuman haploid ES cells. By contrast, the X:A imbalance in differentiating human haploid ES cells is not critical for their survival, and a haploid karyotype is observed in differentiated cells.[17] Hence, unlike nonhuman haploid ES cells, haploid karyotype is not a barrier to human ES cell differentiation.

Spontaneous diploidization can also occur as a result of abnormal cell cycle regulation in haploid cells.[21] Takahashi *etal.* suggested that cell division could be skipped once, by the inclusion of an extra G1/S phase in haploid ES cells, resulting in diploidization. The *Wee1* kinase inhibitor, which accelerates G2/M phase transition and prevents entry into the additional G1/S phase, can be used to improve maintenance of haploidy during long-term in vitro culture.

4. GENOME EDITING

Haploid ES cells can be used in both forward and reverse genetic screens. For example, reverse genetic screens using retroviral mutagenesis have been applied to haploid ES cells, and forward genetic screens were used to identify Gpr107 as a key molecule involved in ricin toxicity.[10] These experiments may be optimized by the application of genome-editing technologies.

4.1 CRISPR/Cas System

The clustered regularly interspaced short palindromic repeats (CRISPR)/CRISPR-associated (Cas) system is a novel genome-editing technology that can be applied in plants and animals, including mammalian cells.[22–25] CRISPR/Cas system consists of the Cas9 endonuclease and a target-specific guide RNA (gRNA) complementary to the target DNA sequence.[26] First, the Cas9 and gRNA complex generate sequence-specific double-strand breaks (DSBs) at target DNA loci bound by gRNAs, and the DSBs are then repaired by one of two alternative methods: nonhomologous end joining (NHEJ) or homology-directed repair (HDR). The repair of DSBs by NHEJ frequently introduces mutations; therefore, the CRISPR/Cas system has been used to perform high-efficiency loss-of-function screens in mouse and human cell lines.[27–29] In addition, the CRISPR/Cas system has been used for knockout and knockin of genes in human-induced pluripotent stem (iPS) cells.[25] This system has been applied for the generation of various disease model cell lines,[30,31] and precise correction of mutant genes mediated by HDR[32–36] in human iPS cells.

4.2 Genome Editing in Haploid ES Cells

We previously reported the application of the CRISPR/Cas system to knockout single or multiple genes in mouse haploid ES cells.[37] For multiple gene knockouts, haploid ES cells were cotransfected with vectors expressing human codon-optimized Cas9 nuclease and gRNAs targeting the ten-eleven translocation 1 (*Tet1*), *Tet2*, and *Tet3* genes. Consequently, all three genes were disrupted at high frequency and corresponding loss-of-function was observed in 10 out of 20 clones (50%) (Fig. 3A). Triple knockout efficiency in haploid ES cells was higher than that previously reported using diploid ES cells (20 out of 96 clones, 20.8%)[38] and our diploid data (2 out of 20 clones, 10%). A study using rat haploid ES cells[15] also targeted all three *Tet* genes (*Tet1*, *Tet2*, and *Tet3*) at high efficiency (3 out of 22 clones, 13.6%). Use of pairs of gRNAs on the same chromosome can generate large chromosomal deletions of up to 1 Mb in mammalian diploid cells.[39] In haploid ES cells, we performed cotransfection with vectors expressing Cas9 and sgRNAs targeting two loci (exons 4 and 7 of the *Tet1* gene).[37] This led to a 14 kb chromosomal deletion (30%) or large inversion (10%) in substantial

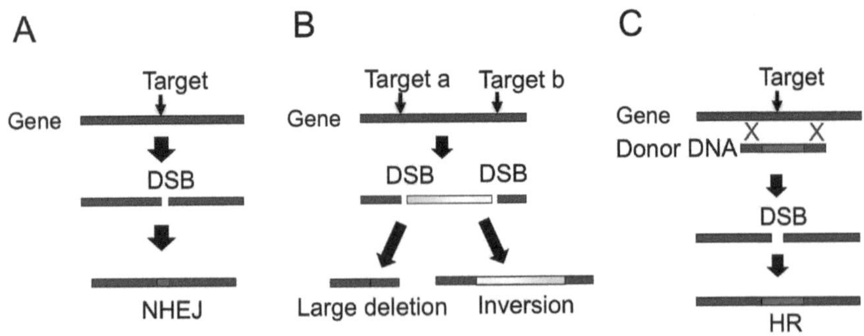

Fig. 3 Genome editing using the CRISPR/Cas system in haploid ES cells. (A) Gene knockout via non-homologous end joining (NHEJ). (B) Large deletion or inversion. (C) Knock-in by homologous recombination (HR). *DSB*, Double-strand break.

proportions of transfected cells (Fig. 3B). In addition, successful CRISPR/Cas-mediated knockin was reported in mouse haploid ES cells (Fig. 3C).[40,41] Kimura *et al.* generated haploid ES cell lines by inserting the P2A-Venus reporter cassette into the *Actb* locus.[40] Notably, the knock-in efficiency was significantly higher in haploid ES cells than in diploid ES cells. This phenomenon may be interpreted in two different ways. First, DSBs induced by Cas9 occurring adjacent to the *Actb* coding sequence may be repaired by the alternative DNA repair systems, NHEJ and HDR, which may lead to disruption of *Actb* function when NHEJ is defective. Hemizygotic haploid ES cells that have lost *Actb* function will die more frequently than diploid ES cells, leading to a reduced number of surviving colonies and an apparent increase in knockin efficiency in haploid ES cells. Second, HDR using information from the homologous chromosome will occur rarely in hemizygotic haploid ES cells, which may increase the opportunity for HDR with exogenous DNA templates.

4.3 Off-Target Effects

Nonspecific recognition and digestion at nontarget regions, so-called "off-target effects," can occur in genome editing, including using the CRISPR/Cas system. As haploid cells contain only one set of chromosomes, disruption of a single allele can easily generate a loss-of-function phenotype. This feature appears to be an advantage of haploid cells; however, disruption of one allele at an off-target site could also generate an undesirable loss-of-function phenotype. Off-target effects have frequently been observed in several human tumor cell lines,[42–44] although they are of low frequency in mouse zygotes and their derivative ES cell lines.[38,45] Nevertheless, off-target effects will occur in any cell line, and to reduce them in haploid cells, it is important to improve target-specific digestion by gRNA and Cas9. For example, the double-nicking method using Cas9 nickase (which cleaves only single-stranded DNA) and a pair of gRNAs can be effective.[46] Notably, double nicking with the paired Cas9 nickase resulted in higher cleavage efficiency than with the unpaired wild-type Cas9 in haploid ES cells.[40] "Enhanced specificity" SpCas9 (eSpCas9), generated by structure-guided mutagenesis, also improves target specificity by maintaining robust on-target cleavage.[47] It is also important to inactivate Cas9 protein after working on target sites; however, the Cas9 expression vector will continue to generate Cas9 protein over a long period of time, resulting in an increase in the incidence of off-target effects. This problem can be avoided by using a recombinant Cas9 protein or a Cas9 inhibitor (anti-Cas9 protein).

5. HAPLOID CELLS AS GAMETE REPLACEMENTS

Genomic imprinting, which is the epigenetic phenomenon by which certain genes are expressed in a parent-of-origin-specific manner (Fig. 4A), is partially maintained in androgenetic and parthenogenetic haploid ES cells; hence, the nuclei of androgenetic and parthogenetic ES cells could be used as sperm[12,13] and oocyte[14] substitutes, respectively (Fig. 4B). Semicloned mice containing heterozygous genetic modifications can be directly produced by nuclear transplantation of genetically manipulated haploid ES cells into oocytes. This can be an effective technique for the introduction of organism-wide mutations in mice in a single generation; however, the birth rate of these semicloned mice was approximately 5% of transferred embryos, and roughly half of live-born semicloned pups were developmentally impaired.[12,13] This may be due to alterations in imprinting and activation of *H19* and *Gtl2* imprinted genes, which occur during derivation and in vitro culture of haploid ES cells. To overcome this, Zhong *et al.* demonstrated that

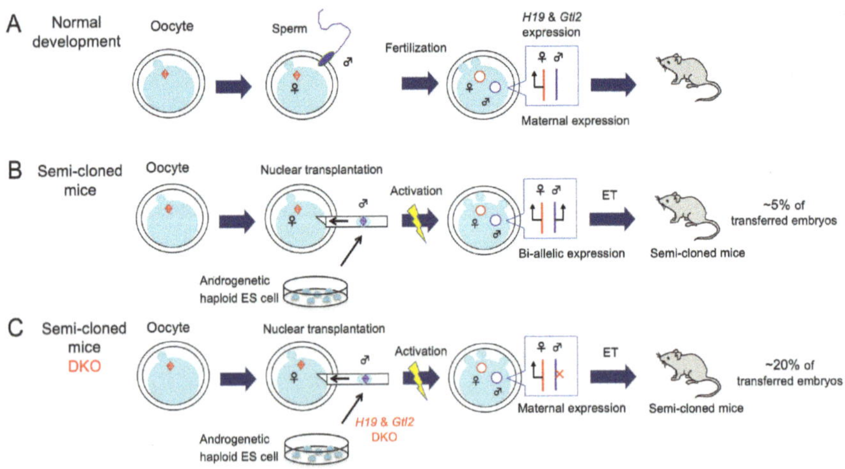

Fig. 4 Generation of semicloned mice using haploid ES cells. (A) Schematic diagram of normal development of mice. *H19* and *Gtl2* genes are expressed from the maternal allele. (B) Schematic diagram of generation of semicloned mice using androgenetic haploid ES cells. *H19* and *Gtl2* genes exhibit biallelic expression due to activation of the allele from haploid ES cells. (C) Schematic diagram of efficient generation of semicloned mice using androgenetic haploid ES cells carrying the *H19* and *Gtl2* double knockout (DKO) allele. The downregulation of *H19* and *Gtl2* improved the development of semicloned mice. *ET*, Embryo transfer.

androgenetic haploid ES cells carrying deletions of the *H19* and *Gtl2* differentially methylated regions (DMRs) could efficiently support the generation of semicloned pups (~20% of transferred embryos) (Fig. 4C).[41]

More surprisingly, mouse–rat allodiploid ES cells can be established by fusing mouse parthenogenetic and rat androgenetic haploid ES cells.[48] The resulting ES cells have stable allodiploid genomes and are capable of differentiating into all three germ layers, both in vivo and in vitro. This technique served as a powerful tool to identify genes escaping X inactivation, and to screen for genes regulated in a species-specific manner, such as those involved in maintenance of pluripotency.

6. GENETIC SCREENING USING HAPLOID ES CELLS

Another application for which haploid ES cells are superior is genome-wide mutational screening, as screening on this scale using diploid cells generally results in ambiguous phenotypes, due to the presence of a second gene copy. In fact, forward genetic screens have been successfully applied in mice to identify factors involved in the DNA mismatch repair pathway and ricin toxicity, by transfection of the gene trap piggyback transposon system into parthenogenetic haploid ES cells.[10,11] Furthermore, a similar strategy has been used in rat,[15] monkey,[16] and human haploid ES cells.[17] Mouse parthenogenetic haploid ES cells have also been used for high-throughput reverse genetics.[10]

Major benefits of haploid cells lie in the ease of genomic manipulation and observation of phenotypes. Pluripotent (or multipotent) haploid ES cells can be differentiated both in vivo and in vitro, facilitating characterization screens of ES cell lines.[27] Near-haploid human tumor cell lines (e.g., KBM-7)[49] have been used for many years in genetic screens focused on identification of the molecular mechanisms underlying reactions to infections and toxins. The recent successful generation of human haploid ES cells[17] will facilitate various analyses using tissues and organs derived from screened ES cell lines.

7. CONCLUDING REMARKS

The combination of the CRISPR/Cas genome-editing system and haploid ES cells allows efficient manipulation of the mammalian genome and provides a new platform for genetic analyses of complex biological phenomena and diseases, such as those that employ forward and reverse genetic screens.

REFERENCES

1. Mortensen RM, Zubiaur M, Neer EJ, Seidman JG. Embryonic stem cells lacking a functional inhibitory G-protein subunit (alpha i2) produced by gene targeting of both alleles. *Proc Natl Acad Sci USA*. 1991;88:7036–7040.
2. Milstone DS, Bradwin G, Mortensen RM. Simultaneous Cre catalyzed recombination of two alleles to restore neomycin sensitivity and facilitate homozygous mutations. *Nucleic Acids Res*. 1999;27:e10.
3. Freed JJ, Mezger-Freed L. Stable haploid cultured cell lines from frog embryos. *Proc Natl Acad Sci USA*. 1970;65:337–344.
4. Yi M, Hong N, Hong Y. Generation of medaka fish haploid embryonic stem cells. *Science*. 2009;326:430–433.
5. Graham CF. The effect of cell size and DNA content on the cellular regulation of DNA synthesis in haploid and diploid embryos. *Exp Cell Res*. 1966;43:13–19.
6. Modlinski JA. Haploid mouse embryos obtained by microsurgical removal of one pronucleus. *J Embryol Exp Morphol*. 1975;33:897–905.
7. Tarkowski AK, Rossant J. Haploid mouse blastocysts developed from bisected zygotes. *Nature*. 1976;259:663–665.
8. Kaufman MH. Chromosome analysis of early postimplantation presumptive haploid parthenogenetic mouse embryos. *J Embryol Exp Morphol*. 1978;45:85–91.
9. Kaufman MH, Robertson EJ, Handyside AH, Evans MJ. Establishment of pluripotential cell-lines from haploid mouse embryos. *J Embryol Exp Morph*. 1983;73:249–261.
10. Elling U, Taubenschmid J, Wirnsberger G, et al. Forward and reverse genetics through derivation of haploid mouse embryonic stem cells. *Cell Stem Cell*. 2011;9:563–574.
11. Leeb M, Wutz A. Derivation of haploid embryonic stem cells from mouse embryos. *Nature*. 2011;479:131–134.
12. Yang H, Shi L, Wang BA, et al. Generation of genetically modified mice by oocyte injection of androgenetic haploid embryonic stem cells. *Cell*. 2012;149:605–617.
13. Li W, Shuai L, Wan H, et al. Androgenetic haploid embryonic stem cells produce live transgenic mice. *Nature*. 2012;490:407–411.
14. Wan H, He Z, Dong M, et al. Parthenogenetic haploid embryonic stem cells produce fertile mice. *Cell Res*. 2013;23:1330–1333.
15. Li W, Li X, Li T, et al. Genetic modification and screening in rat using haploid embryonic stem cells. *Cell Stem Cell*. 2014;14:404–414.
16. Yang H, Liu Z, Ma Y, et al. Generation of haploid embryonic stem cells from *Macaca fascicularis* monkey parthenotes. *Cell Res*. 2013;23:1187–1200.
17. Sagi I, Chia G, Golan-Lev T, Peretz M, Weissbein U, Sui L, Sauer MV, Yanuka O, Egli D, Benvenisty N. Derivation and differentiation of haploid human embryonic stem cells. *Nature*. 2016;532:107–111.
18. Leeb M, Walker R, Mansfield B, Nichols J, Smith A, Wutz A. Germline potential of parthenogenetic haploid mouse embryonic stem cells. *Development*. 2012;139: 3301–3305.
19. Leeb M, Perry AC, Wutz A. Establishment and use of mouse haploid ES cells. *Curr Protoc Mouse Biol*. 2015;5:155–185.
20. Horii T, Hatada I. Genome editing using mammalian haploid cells. *Int J Mol Sci*. 2015;16:23604–23614.
21. Takahashi S, Lee J, Kohda T, Matsuzawa A, Kawasumi M, Kanai-Azuma M, Kaneko-Ishino T, Ishino F. Induction of the G2/M transition stabilizes haploid embryonic stem cells. *Development*. 2014;141:3842–3847.
22. Cho SW, Kim S, Kim JM, Kim JS. Targeted genome engineering in human cells with the Cas9 RNA-guided endonuclease. *Nat Biotechnol*. 2013;31:230–232.
23. Cong L, Ran FA, Cox D, et al. Multiplex genome engineering using CRISPR/Cas systems. *Science*. 2013;339:819–823.

24. Jinek M, East A, Cheng A, Lin S, Ma E, Doudna J. RNA-programmed genome editing in human cells. *Elife.* 2013;2:e00471.
25. Mali P, Yang L, Esvelt KM, Aach J, Guell M, DiCarlo JE, Norville JE, Church GM. RNA-guided human genome engineering via Cas9. *Science.* 2013;339:823–826.
26. Jinek M, Chylinski K, Fonfara I, Hauer M, Doudna JA, Charpentier E. A programmable dual-RNA-guided DNA endonuclease in adaptive bacterial immunity. *Science.* 2012;337:816–821.
27. Koike-Yusa H, Li Y, Tan EP, Velasco-Herrera Mdel C, Yusa K. Genome-wide recessive genetic screening in mammalian cells with a lentiviral CRISPR-guide RNA library. *Nat Biotechnol.* 2014;32:267–273.
28. Shalem O, Sanjana NE, Hartenian E, et al. Genome-scale CRISPR-Cas9 knockout screening in human cells. *Science.* 2014;343:84–87.
29. Wang T, Wei JJ, Sabatini DM, Lander ES. Genetic screens in human cells using the CRISPR-Cas9 system. *Science.* 2014;343:80–84.
30. Horii T, Tamura D, Morita S, Kimura M, Hatada I. Generation of an ICF syndrome model by efficient genome editing of human induced pluripotent stem cells using the CRISPR system. *Int J Mol Sci.* 2013;14:19774–19781.
31. An MC, O'Brien RN, Zhang N, Patra BN, De La Cruz M, Ray A, Ellerby LM. Polyglutamine disease modeling: epitope based screen for homologous recombination using CRISPR/Cas9 system. *PLoS Curr.* 2014;6.
32. Xie F, Ye L, Chang JC, Beyer AI, Wang J, Muench MO, Kan YW. Seamless gene correction of β-thalassemia mutations in patient-specific iPSCs using CRISPR/Cas9 and piggyback. *Genome Res.* 2014;24:1526–1533.
33. Li HL, Fujimoto N, Sasakawa N, et al. Precise correction of the dystrophin gene in Duchenne muscular dystrophy patient induced pluripotent stem cells by TALEN and CRISPR-Cas9. *Stem Cell Rep.* 2015;4:143–154.
34. Ousterout DG, Kabadi AM, Thakore PI, Majoros WH, Reddy TE, Gersbach CA. Multiplex CRISPR/Cas9-based genome editing for correction of dystrophin mutations that cause Duchenne muscular dystrophy. *Nat Commun.* 2015;6:6244.
35. Song B, Fan Y, He W, Zhu D, Niu X, Wang D, Ou Z, Luo M, Sun X. Improved hematopoietic differentiation efficiency of gene-corrected beta-thalassemia induced pluripotent stem cells by CRISPR/Cas9 system. *Stem Cells Dev.* 2015;24: 1053–1065.
36. Huang X, Wang Y, Yan W, Smith C, Ye Z, Wang J, Gao Y, Mendelsohn L, Cheng L. Production of gene-corrected adult beta globin protein in human erythrocytes differentiated from patient iPSCs after genome editing of the sickle point mutation. *Stem Cells.* 2015;33:1470–1479.
37. Horii T, Morita S, Kimura M, et al. Genome engineering of mammalian haploid embryonic stem cells using the Cas9/RNA system. *PeerJ.* 2013;1:e230.
38. Wang H, Yang H, Shivalila CS, Dawlaty MM, Cheng AW, Zhang F, Jaenisch R. One-step generation of mice carrying mutations in multiple genes by CRISPR/Cas-mediated genome engineering. *Cell.* 2013;153:910–918.
39. Canver MC, Bauer DE, Dass A, Yien YY, Chung J, Masuda T, Maeda T, Paw BH, Orkin SH. Characterization of genomic deletion efficiency mediated by clustered regularly interspaced palindromic repeats (CRISPR)/Cas9 nuclease system in mammalian cells. *J Biol Chem.* 2014;289:21312–21324.
40. Kimura Y, Oda M, Nakatani T, Sekita Y, Monfort A, Wutz A, Mochizuki H, Nakano T. CRISPR/Cas9-mediated reporter knock-in in mouse haploid embryonic stem cells. *Sci Rep.* 2015;5:10710.
41. Zhong C, Yin Q, Xie Z, et al. CRISPR-Cas9-mediated genetic screening in mice with haploid embryonic stem cells carrying a guide RNA library. *Cell Stem Cell.* 2015;17: 221–232.

42. Fu Y, Foden JA, Khayter C, Maeder ML, Reyon D, Joung JK, Sander JD. High-frequency off-target mutagenesis induced by CRISPR-Cas nucleases in human cells. *Nat Biotechnol.* 2013;31:822–826.
43. Hsu PD, Scott DA, Weinstein JA, et al. DNA targeting specificity of RNA-guided Cas9 nucleases. *Nat Biotechnol.* 2013;31:827–832.
44. Pattanayak V, Lin S, Guilinger JP, Ma E, Doudna JA, Liu DR. High-throughput profiling of off-target DNA cleavage reveals RNA-programmed Cas9 nuclease specificity. *Nat Biotechnol.* 2013;31:839–843.
45. Yang H, Wang H, Shivalila CS, Cheng AW, Shi L, Jaenisch R. One-step generation of mice carrying reporter and conditional alleles by CRISPR/Cas-mediated genome engineering. *Cell.* 2013;154:1370–1379.
46. Ran FA, Hsu PD, Lin CY, et al. Double nicking by RNA-guided CRISPR Cas9 for enhanced genome editing specificity. *Cell.* 2013;154:1380–1389.
47. Slaymaker IM, Gao L, Zetsche B, Scott DA, Yan WX, Zhang F. Rationally engineered Cas9 nucleases with improved specificity. *Science.* 2015;351(6268):84–88.
48. Li X, Cui XL, Wang JQ, Wang YK, Li YF, Wang LY, Wan HF, Li TD, Feng GH, Shuai L, Li ZK, Gu Q, Hao J, Wang L, Zhao XY, Liu ZH, Wang XJ, Li W, Zhou Q. Generation and application of mouse-rat allodiploid embryonic stem cells. *Cell.* 2016;164:279–292.
49. Kotecki M, Reddy PS, Cochran BH. Isolation and characterization of a near-haploid human cell line. *Exp Cell Res.* 1999;252:273–280.

FURTHER READING

Evans MJ, Kaufman MH. Establishment in culture of pluripotential cells from mouse embryos. *Nature.* 1981;292:154–156.

Guo G, Yang J, Nichols J, Hall JS, Eyres I, Mansfield W, Smith A. Klf4 reverts developmentally programmed restriction of ground state pluripotency. *Development.* 2009;136:1063–1069.

CHAPTER SIX

CRISPR in Animals and Animal Models

Ellen Shrock*,†, Marc Güell†,‡,*

[*] Biological and Biomedical Sciences, Harvard University, Boston, MA, United States
[†] Harvard Medical School, Boston, MA, United States
[‡] Pompeu Fabra University, Barcelona, Spain
[1] Corresponding author. E-mail address: marcguellc@gmail.com

Contents

Abstract

CRISPR–Cas9 has revolutionized the generation of transgenic animals. This system has demonstrated an unprecedented efficiency, multiplexability, and ease of use, thereby reducing the time and cost required for genome editing and enabling the production of animals with more extensive genetic modifications. It has also been shown to be applicable to a wide variety of animals, from early-branching metazoans to primates. Genome-wide screens in model organisms have been performed, accurate models of human diseases have been constructed, and potential therapies

Progress in Molecular Biology and Translational Science, Volume 152
ISSN 1877-1173
http://dx.doi.org/10.1016/bs.pmbts.2017.07.010

have been tested and validated in animal models. Several achievements in genetic modification of animals have been translated into products for the agricultural and pharmaceutical industries. Based on the remarkable progress to date, one may anticipate that in the future, CRISPR–Cas9 technology will enable additional far-reaching advances, including understanding the bases of diseases with complex genetic origins, engineering animals to produce organs for human transplantation, and genetically transforming entire populations of organisms to prevent the spread of disease.

1. INTRODUCTION

1.1 CRISPR–Cas9 Genome Editing Technology

Genome editing technologies have had a profound impact on fundamental, clinical, and industrial research involving transgenic animals. Before the advent of the RNA-guided endonuclease CRISPR–Cas9 (Clustered Regularly Interspaced Short Palindromic Repeats-CRISPR-associated protein 9), genome editing in animals was performed by the use of chemical and UV mutagenesis, DNA recombinase-mediated gene replacement, zinc-finger nucleases (ZFNs) and transcription activator-like effector nucleases (TALENs).[1] CRISPR–Cas9 has proven to be a much simpler and more efficient method, and for this reason it has revolutionized the field of genome editing and enabled modifications that would have been unfeasible or impractical using prior technologies.[2–5]

The CRISPR–Cas9 system is an adaptive immune system used by bacteria to protect against invading viruses and plasmids through sequence-specific cleavage of these foreign nucleic acids.[6,7] As repurposed for genome editing, it consists of a ~20 nucleotide single guide RNA (sgRNA) that undergoes Watson–Crick base-pairing with a target DNA sequence preceding an NGG (where N denotes any of the four DNA bases) protospacer adjacent motif, complexed with an endonuclease, Cas9, that carries out a DNA double-stranded break after the sgRNA has bound to the target DNA sequence.[8–10] Unlike ZFNs or TALENs, which require the design of a protein to target a specific DNA locus, CRISPR–Cas9 only requires the much simpler task of designing a ~20 nucleotide sgRNA, which determines the site of DNA cleavage.

The CRISPR–Cas9 system is used as a tool to inactivate or modify genes via sequence-specific double-strand breaks (DSBs). Cas9-induced DSBs are recognized by the cellular DNA damage response machinery and can be repaired by endogenous DSB repair pathways. The predominant repair pathway is nonhomologous end joining (NHEJ), which often results in small insertions and/or deletions (indels) that can create frameshift mutations and disrupt the function of genes. This pathway can be exploited to generate genetic knockout mutations. Alternatively, in the presence of repair templates, that is, genomic or exogenous DNA with homology to the site of the DSB, the damage can be repaired seamlessly by homology-directed repair (HDR). HDR can be exploited to introduce specific sequence alterations to endogenous genes.[5]

Although CRISPR–Cas9 technology has been very successful, it does have some limitations. HDR-mediated genome editing to introduce precise genetic modifications remains a much less efficient process than NHEJ-mediated gene disruption, though progress has been made in this regard.[11–13] Multi-kb replacements by the HDR pathways are challenging and require selections and/or large population cell sorting.[14] Consequently, the major applications for the HDR pathways are the local replacement of key regions within genes. The ability to conveniently introduce precise site-specific modifications of unlimited size currently represents an unmet need. Thus, CRISPR–Cas9-mediated modifications are largely limited to applications involving gene disruption, applications where modified cells have a competitive advantage over wild-type cells, and/or applications where low-efficiency genetic modification is sufficient to produce the desired phenotype. A second limitation relates to the off-target effects of Cas9, which raise concerns about inadvertently disrupting genes of physiological importance[15,16] or causing chromosomal deletions, inversions, or translocations.[17] One can address these concerns with a number of strategies to increase CRISPR–Cas9 specificity and decrease off-target activity, including (i) selecting sgRNAs with few off-target sites within the genome using bioinformatics software,[18] (ii) transiently delivering Cas9 mRNA or ribonucleoproteins (RNPs),[19] (iii) using sgRNA or Cas9 variants with reduced DNA binding activity, such as truncated, 17–18 nt sgRNAs,[20] high-fidelity Cas9 (Cas9-HF1),[21] or enhanced-specificity Cas9 (eCas9),[22] and (iv) using multiple gRNAs together with paired nickases[23] or fokI fusions to inactivated Cas9 (dCas9).[24] A third limitation arises from the delivery requirement and the fact that the commonly used Cas9 from *Streptococcus pyogenes* bacteria (SpCas9) is approximately 4 kb and hence is a rather large cargo for a viral vector, such as adeno-associated virus (AAV). The MIT lab of Zhang and coworkers has developed another variant of Cas9, derived from the *Staphlococcus aureus* bacteria (SaCas9) that is 1 kb in length—substantially shorter than SpCas9, thus expediting delivery with AAV.

This group has reported that the SaCas9 exhibits an efficiency in genome editing that is similar to that of SpCas9.[25] In the past, most methods of genetic modification were restricted to use in model organisms, such as the mouse, whose cells could be manipulated via homologous recombination.[1] CRISPR–Cas9 has widened the scope of transgenic animal production by enabling facile genetic modification of embryonic stem cells (ESCs), as well as direct modification of zygotes, which is applicable for most species.[26] CRISPR–Cas9 has been used productively in a wide variety of animals, including nematodes,[27] fruit flies,[28] malarial mosquitoes,[29] sea urchins,[30] zebrafish,[31] sea anemone,[32] mice,[33] nonhuman primates (NHPs),[34] and even human zygotes.[35] These examples will be discussed below.

2. METHODS OF PRODUCING GENETICALLY MODIFIED ANIMALS

The CRISPR–Cas9 system has been used to generate a wide variety of transgenic animals via a number of common methods. These include: (i) germline editing via injection or electroporation of CRISPR–Cas9 reagents into early-stage zygotes or gonads, often used in model organisms, such as *Caenorhabtidis elegans, Danio rerio,* and *Drosophila melanogaster,* as well as in larger animals, (ii) primary cell editing followed by somatic cell nuclear transfer (SCNT), often used in larger animals, such as pigs and cows, and (iii) blastocyst injection with genetically modified ESCs followed by breeding to produce stable germline mutants, traditionally used in mice (Fig. 1). CRISPR–Cas9 reagents can be introduced in the form of plasmids, mRNA preparations, as RNPs, or in viral vectors such as AAVs.

2.1 Small Model Organisms

In *C. elegans*, expression of Cas9 and gRNA can easily be carried out by injecting an expression plasmid, mRNA, or ribonuclear protein complexes into the gonads of the worm.[36] Single worm polymerase chain reaction (PCR) is commonly used to evaluate the outcome of the gene editing experiment. In this procedure, multiple F1 worms are placed in new plates, allowed to generate eggs, and then processed with PCR to find heterozygotes. The F2 progeny of positive F1's are used to identify homozygotes. In the fruit fly *Drosophila,* CRISPR–Cas9 can be efficiently used to generate germline genetic modifications. CRISPR–Cas9 reagents can be injected into preblastoderm embryos, and cells from these embryos can be screened by PCR 24 h

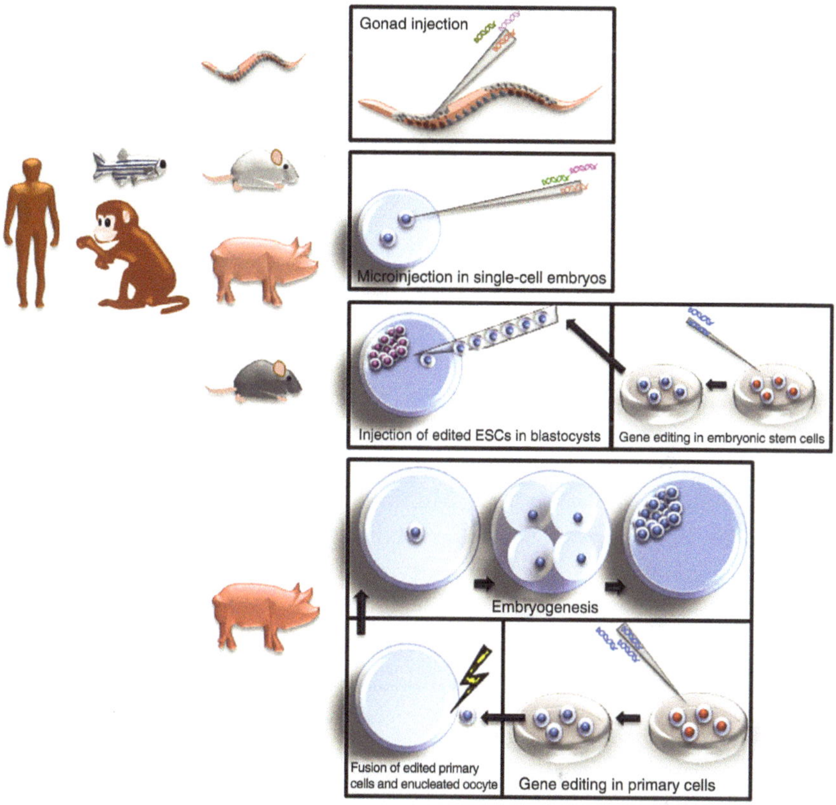

Fig. 1 Methods of producing transgenic animals. Common methods of producing transgenic animals include: microinjection of gene editing reagents into the gonad syncytium, microinjection of gene editing reagents into the pronuclei of fertilized oocytes, microinjection of edited embryonic stem cells (ESCs) into the blastocyst, and somatic cell nuclear transfer (SCNT).

after injection. Pairs of gRNAs have been used successfully to generate defined deletions, and, in combination with single-strand oligo-deoxynucleotides, to induce Cas9-mediated homologous recombination.[28] This method has also been widely used with zebrafish. CRISPR–Cas9 reagents are injected into zygotes, or early-stage embryos to generate mosaic F0 animals. Mosaic adults are then crossed to generate homozygous mutants.[31,37,38]

2.2 Rodents

The production, via conventional means, of mice carrying mutations in multiple genes was previously a labor-intensive and slow process involving sequential recombination in ESCs, injection of modified ESCs into blastocysts, and interbreeding of singly mutated mice. This method of transgenesis has been accelerated greatly with the advent of CRISPR–Cas9. A major advance was reported by Jaenisch and coworkers,[33] in which they successfully used CRISPR–Cas9 in a one-step process to produce mice with mutations in multiple genes. This group demonstrated precise genome editing by coinjecting fertilized oocytes at the pronuclear stage with not just the Cas9 mRNA and sgRNA, but also single-stranded DNA oligos, enabling HDR based on the DNA oligos as templates. Blastocysts with these DNA oligo injections were implanted into the wombs of foster mothers. The time scale required for the generation of mice carrying multiple mutations using pronuclear injection was shown to be approximately one month, much shorter than the six-month to year-long time scale required by conventional methods. In addition to germline editing, ex vivo and in vivo gene editing has been demonstrated. Multiple systems have been developed to perform in vivo CRISPR–Cas9 editing in rodents. The necessary reagents are generally packaged in lentiviruses or AAV. Multiple tissues, such as lung,[39] neurons,[40] or liver[41] have been targeted. Delivery strategies using cationic nanoparticles have also been developed.[42]

2.3 Larger Animals

The production of larger transgenic mammals, including pigs[43] and cows,[44] has been accomplished via SCNT and/or pronuclear injection. SCNT involves genetically engineering primary cell lines in vitro, then fusing the nuclei of individual cells with enucleated porcine oocytes to produce genetically modified embryos.[45] At the blastocyst stage of development, these embryos are delivered to the uterus of a surrogate mother. Transgenic pigs may also be generated using zygote injection of the CRISPR–Cas9 reagents.[46]

2.4 Nonhuman Primates

Although NHPs require larger facilities and longer time periods for experiments as compared with mice, they nonetheless play an important role in biomedical research[47] as they are phylogenetically closer to humans. The CRISPR–Cas9 system has been of great value for genome editing in NHPs. Indeed, highly precise genetic modifications in monkeys had not been feasible before the advent of CRISPR–Cas9, as the only technologies available for gene editing were viral vectors.[48–50] CRISPR–Cas9 has been used to precisely edit the genome of the Cynomolgus monkey. Using zygote injection, researchers introduced Cas9 mRNA and five gRNAs targeting the *Nr0b1*, *PPar-gamma*, and *Rag1* genes.[34] In the future, it is thought that genetically engineered NHPs may be very useful to study neurological disorders and human traits that cannot be easily recapitulated in common model organisms.

3. USES OF GENETICALLY MODIFIED ANIMALS

The uses of CRISPR–Cas9 for animal genome engineering include several specific applications: (i) the inactivation or alteration of genes in model animals in order to elucidate the functions of these genes; (ii) the production of animal models of human disease to study disease progression in a controlled manner and evaluate potential therapies; (iii) and the use of genetically modified animals for industrial, pharmaceutical, and biotechnological production Fig. 2.

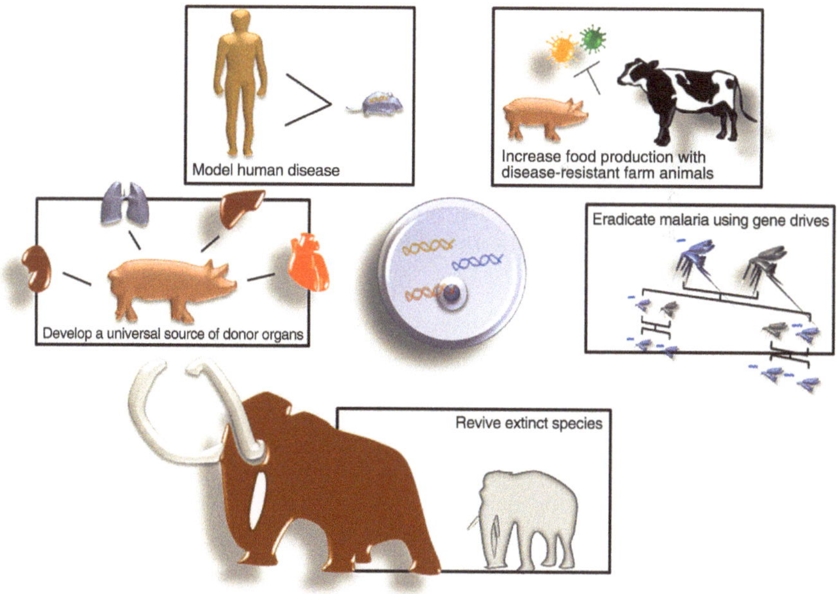

Fig. 2 Utilizations of genetically modified organisms. Present and future applications of genetically engineered animals include modeling human disease, increasing agricultural production with disease-resistant animals, developing pigs whose organs may be used for human transplantation, eradicating malaria with mosquitoes carrying gene drives, and reviving extinct species.

3.1 Basic Scientific Research

A fundamental question in biology concerns the way in which genotypes are translated into phenotypes. As genome sequencing technologies have advanced, there have been corresponding improvements in understanding genotype–phenotype relationships. Among these, the identification of human disease–causing mutations have been of particular importance. The production of transgenic animals with candidate genetic variants has been an important method for validating and studying pathologies resulting from these variants. Here again, CRISPR–Cas9 has been of great value.

3.1.1 Rapid Reverse CRISPR Screening

One of the most remarkable features of the CRISPR–Cas9 system is its facile programmability, which allows for a high capacity for multiplexing, that is, targeting multiple genomic loci. As a result, CRISPR–Cas9 technology has quickly been adopted for pooled sgRNA screens targeting up to thousands of genomic loci simultaneously in cell lines.[51,52] This has greatly facilitated the determination of gene function and has made possible genome-wide screens for essential genes. These and other CRISPR screens have been carried out in vivo, using zebrafish.[53,54] Recently, a CRISPR screen in mice was used to identify potential targets for cancer therapies.[55,56] In vivo CRISPR screens have proven to be a powerful platform to conduct reverse genetics to identify genes involved in vertebrate biological processes and identify candidate genes for new therapies for human diseases.

3.2 Animal Models of Human Diseases

Over the last years more and more information has been obtained linking genotypes with disease phenotypes. However, the pace at which these pathogenic alleles can be tested has been limited. CRISPR–Cas9 genome editing has greatly expedited the process of studying disease-causing mutations in animal models.[26]

Single nucleotide polymorphisms (SNPs) cause various diseases. Introducing small genome modifications to recapitulate these SNPs in animal models can be performed by introducing a gRNA that targets a position near to the target locus, in conjunction with Cas9, and a donor template DNA.[55] Longer deletions or insertions are useful to recapitulate human genome deletions, to replace endogenous genes with human versions, or to modify regulatory elements in animal models. Deletions can be achieved by combining Cas9 with paired gRNA targeting the boundaries of the intended deletion. Deletions of up to 1 Mb in length have been reported in cells using this technique.[57] Insertion of large cassettes of DNA can be achieved by using gRNA, Cas9, and donor DNA containing the intended knockin sequence flanked by homology arms.[58]

3.2.1 Cancer

Mouse models of human genetic traits, development, and diseases have been utilized heavily in basic and applied research.[26,59–61] One particularly important application of disease modeling using CRISPR has been in recapitulating the onset and progression of cancer in mice. To this end, the Zhang and Sharp labs at MIT/Broad Institute used CRISPR–Cas9 both in vivo and ex vivo in neurons, cells of the immune system, and endothelial cells of mice to model the progression of the three most important genes whose mutations lead to lung cancer, namely Kras, Tp53, and Lkb1. These groups were able to deliver CRISPR–Cas9 reagents using a variety of means, including AAV and lentivirus.[62] A MIT-Harvard team[63] used hydrodynamic tail-vein injection to deliver CRISPR–Cas9 reagents to the mouse liver in vivo to target and disrupt the tumor suppressor genes Pten and Tp53 alone and in combination. This team showed that the resultant inactivation of these genes in vivo led to cancerous tumors. This group also used CRISPR–Cas9 to disrupt the β-catenin gene, whose mutations occur in liver cancer, and obtained similar results. A group at Sloan-Kettering used CRISPR–Cas9 technology to engineer oncogenic chromosomal rearrangements in mice.[64] It had been known that such chromosomal rearrangements had been implicated in lung cancer, but it had been challenging to study them with mouse models because of the difficulty and complexity of manipulating the germ line. Again, CRISPR–Cas9 technology was found to greatly expedite and facilitate the study of this oncogenic process. These works demonstrated the valuable role of CRISPR–Cas9 in studying cancer tumor progression. CRISPR–Cas9 was also used to produce myeloid tumors in mouse models.[65]

3.2.2 Genetic Disease and Gene Therapy

Another major advance has been in the use of CRISPR–Cas9 genome editing for gene therapy, in particular, to improve muscle function in mouse models of one of the most common genetic diseases, Duchenne muscular dystrophy (DMD). DMD is a disease resulting from a mutation in both copies of the gene coding for dystrophin, a protein that helps to maintain the membranes of muscle cells. The symptom of this ultimately fatal disease is progressive loss of muscular function, and there is no known cure. Conventional gene therapy attempts had sought to insert wildtype copies of the dystrophin gene in cells via a viral vector, such as AAV, but had been relatively ineffective because of the large size of the dystrophin gene. In late 2015, three labs reported the successful use of CRISPR–Cas9 to partially alleviate muscular dystrophy in mice; these were a collaboration of the Church and Wagers labs at Harvard,[66] the Gersbach lab at Duke,[67,68] and the Olson lab at the Texas Southwestern Medical Center in Dallas.[69] The strategy for the application of CRISPR–Cas9 to gene therapy for DMD was the design of sgRNAs to excise a frame-shifting mutated exon and thereby restore in-frame translation and resultant production of a shortened but still functional dystrophin protein. For this purpose, the mdx mouse model of DMD, which has a frame-shifting mutation in the dystrophin exon 23, was chosen. The sgRNAs were designed to produce double-stranded breaks on either side of this exon 23, thereby removing it. This was delivered by AAV to several types of muscle tissue, including myofibers, cardiomyocytes, and muscle stem cells, as a local or systemic treatment, in young (postnatal) and adult mdx mice. This resulted in the successful production of the shortened but still functional dystrophin in vivo. This was a milestone achievement by these three labs, since it was the first time that the CRISPR–Cas9 system was used for gene therapy in a grown animal. More recently, CRISPR–Cas9 has been used to correct a dystrophin mutation in skeletal muscle stem cells in a mouse model of muscular dystrophy.[70]

3.3 Industrial Production

3.3.1 Use of CRISPR in Farming

Gene editing using CRISPR—Cas9 has impacted industrial production by accelerating the creation of animals with novel genotypes that may be beneficial for agricultural production. For example, the company AgGenetics has engineered Aberdeen Angus cows—the cows that are bred for Angus beef—to produce short white hair, which is more adequate for warmer climates.[71] This could enable Aberdeen Angus herds to thrive in a greater number of environments, and thus increase agricultural yields of Angus beef. The company Recombinetics has produced a hornless breed of cattle using genome editing. This may render it unnecessary to use hot irons to burn off horns in the processing of cattle for food.[72] CRISPR–Cas9 has even been used to produce hypoallergenic eggs. While eggs are used in multiple biotechnological products, such as vaccines, and are of high nutritional value, they contain certain allergens that prevent some individuals from being able to receive vaccines or consume egg products.[73] CRISPR–Cas9 technology has been used to create genetically modified chickens that lack the allergens ovalbumin and ovomucoid and that, as a result, produce hypoallergenic eggs.[74]

Gene editing has been leveraged to produce animals resistant to common pathogens that cause major economic losses in the feedstock industry. CRISPR–Cas9 has been used to generate pigs resistant to the Porcine Reproductive and Respiratory Syndrome Virus (PRRSV), a pathogen that causes large numbers of late-term abortions and stillbirths in pigs. This was accomplished by inactivation of the gene *CD163*, which encodes the porcine receptor for PRRSV.[75] Recently, CRISPR–Cas9 was used to introduce a gene in cattle that provides resistance to bovine tuberculosis.[44] As these examples show, gene editing technology is being used effectively in the food production industry, with important implications for cost-effective agricultural production and animal health. Although this article is limited to applications of CRISPR–Cas9 involving animals, it should be mentioned that this technology has also been of great value for improvements in plant agricultural production.[76]

3.3.2 Pharmaceutical Production

Genetically modified animals have been used in the production of recombinant pharmaceuticals. For example, the company GTC Biotherapeutic engineered goats (using methods other than CRISPR–Cas9) to produce antithrombin and secrete this pharmaceutical agent in their milk. The Pharmin Group has engineered transgenic rabbits to similarly produce and secrete a C1-esterase inhibitor. Although these examples have demonstrated that transgenic animals can be excellent sources of recombinant pharmaceuticals, this strategy has been difficult to generalize due to technical difficulties. It is expected that the CRISPR–Cas9 system will expedite the production of new animals as "bioreactors" for pharmaceutical production.[77]

3.3.3 Domestic Animals

Genetically modifying pets to produce custom traits has been on the agenda of the gene editing field for several years. Very small pigs (often called "micropigs") have been produced using CRISPR–Cas9 and advertised as pets.[78] Plans are underway to use gene editing technology to design custom coat patterns in domestic animals.[79] CRISPR–Cas9 has also been used to produce dogs with improved running ability. This was accomplished by inactivating MSTS, a negative regulator of skeletal muscle mass.[80] In addition to genetically modifying domestic animals to produce custom traits, in the future, animals may be engineered in order to correct pathological recessive mutations accumulated through inbreeding and improve overall animal health.

4. LOOKING FORWARD

Genetic engineering with CRISPR–Cas9 is opening exciting new possibilities beyond engineering animals in the laboratory or on the farm. High-capacity gene editing may make possible the production of human-compatible porcine organs, the revival of extinct genetic traits or organisms, and the introduction of novel genetic technologies such as "gene drives," enabling genetic engineering throughout entire ecosystems (Fig. 2).

4.1 Transgenic Pigs for Xenotransplantation

The shortage of human organs for transplantation is a major barrier to the treatment of patients suffering from organ failure.[81] Although porcine organs are considered promising, their use has been hindered by concerns about the transmission of porcine endogenous retroviruses (PERVs) to humans. Moreover, their use in pig-to-human xenotransplantation has been stymied by previously insurmountable immunological incompatibilities.[82–84] The potential of CRISPR–Cas9 gene editing to eliminate the risk of PERV transmission has been realized:[85] in a breakthrough study in 2015, members of the Church lab at Harvard Medical School demonstrated the successful use of CRISPR–Cas9 to inactivate, for the first time, all 62 copies of PERVs in a porcine cell line.[86] This work set a record for the number of loci modified in a single CRISPR–Cas9 experiment. In 2017, these authors further announced the production of healthy, PERV-free pigs whose organs would be used in preclinical trials of xenotransplantation.[87] Current research involves the use of CRISPR–Cas9 to improve immunocompatibility between porcine organs and humans, as has previously been demonstrated between porcine organs and NHPs (baboons).[88]

Perhaps in the future, gene editing will also make possible the production of human tissues in other species via chimeric interspecies blastocyst complementation. In pioneering work by Kobayashi *et al.*,[89] researchers injected wild-type rat pluripotent stems cells in Pdx1-null (pancreatogenesis essential factor) mouse blastocysts, generating a normally-functioning rat pancreas in Pdx1-null mice. In early 2017, Wu *et al.* reported the use of CRISPR–Cas9 to generate rat–moue chimeras by combining zygote-injection of Cas9/gRNA targeting Pdx1 and blastocyst complementation with rat pluripotent stem cells. These authors also tested other chimeras, such as pig–human and pig–rat combinations. The study confirmed a large degree of chimerism in rat–mouse and a more limited degree of chimerism in human–pig combinations.[90]

4.2 Ecological Engineering

CRISPR–Cas9-based gene drives have been proposed as a tool to edit genomes at the population level through propagation of genetic traits by super-Mendelian heritage. Such engineering may help to control disease transmission and protect ecosystems from invasive species.[91] At the molecular level, CRISPR–Cas9-based gene drives operate by cleaving the wild type version of a gene and promoting the cell to repair by copying the homologous version of the gene—which contains the gene drive—to the damaged version. In this way, heterozygote carriers of a gene drive are transformed into homozygotes and the gene drive is passed on to all progeny. The progeny are likewise transformed into homozygotes, and this cycle continues until the gene drive spreads throughout the population. Gene drives have been proposed as a potential mechanism to prevent the spread of malaria via mosquitoes. The efficacy of gene drives for spreading traits has been demonstrated in *Drosophila melanogaster*,[92–95] Gene drives that produce a mutation causing sterility have recently been tested in two species of mosquitoes that transmit malaria, namely *Anopheles gambiae*[96] and *Anopheles stephensi*.[97] Using a CRISPR–Cas9 gene drive, a group at the Imperial College in London demonstrated mutation transmission rates to progeny of 91.4%–99.6% in *A. gambiae*.[29] To the extent that the gene drive spreads the sterility mutation throughout a large fraction of the mosquito population, this could substantially reduce the incidence of malarial infections in humans.

Another application of ecosystem-wide engineering is in the eradication of invasive species. For example, some species of invasive rodents have a negative impact on ecosystems they invade and endanger local fauna.[98] Researchers have experimented in creating mice with a gene from the Y chromosome inserted as a gene drive into chromosome 17 such that births produce only male offspring.[91] One cautionary comment is that resistance against CRISPR–Cas9 gene drives can develop because of insertions and deletions (indels) introduced by NHEJ after an initial Cas9 cleavage event that can mutate the wild type gene sequence and prevent propagation of the drive.[99]

4.3 Restoring Extinct Species

One future application of CRISPR–Cas9 genome engineering is in the revival of extinct animals or, more generally, the revival of traits exhibited by extinct animals.[100,101] The feasibility of this process of "deextinction" depends on the specific species. For recent species where living tissues have been collected and preserved, SCNT could be used with a closely related species as the surrogate mother. In 2000, this interspecies approach was successfully used to "resurrect" a species of goat that had recently become extinct.[102] Even If no living tissues were available, it could still be possible to isolate the DNA of the extinct species, sequence and assemble the genome, perform genome-wide comparisons with the genomes of related species, and then modify the genome of the most closely related species to resemble that of the extinct species.[103] For example, while wooly mammoths and elephants are genetically similar, the genomes of wooly mammoths exhibit variations in genes—including ones related to heat sensing and hair growth—that enabled this species to survive in the arctic tundra. Using CRISPR–Cas9, one could engineer the elephant versions of these genes to resemble the wooly mammoth versions. In addition to enabling the study of ancient alleles that have been lost over the course of evolution, a further purpose could be to revive a species that occupies a certain niche and has a beneficial effect on surrounding ecology. In the case of the mammoth, it is thought that reviving a mammoth-like creature to occupy the arctic tundra could help to slow climate change by reverting the Siberian tundra back towards a grassland landscape.[104]

5. ETHICAL CONSIDERATIONS

There have been, and continue to be, important debates surrounding the ethical aspects of gene editing technologies.[105,106] As CRISPR–Cas9 technology has accelerated the ability to create transgenic animals, these discussions have become increasingly relevant. Debates have been most intense regarding editing of the human genome, especially of the germline, but have also concerned editing of nonhuman animal genomes. For example, the proposal to use CRISPR–Cas9 gene drives to eradicate alleles or entire species from ecosystems has been the subject of ethical consideration. Despite the potential benefits from disease control or elimination of invasive species, the dispersion of synthetic CRISPR–Cas9 systems throughout entire ecosystems may be difficult to control and may have unintended consequences on neighboring species. The subject of another debate has been the proposed development of human/animal chimeras to produce large supplies of human organs for transplantation, as this may involve the risk of human cell contribution in neural or germ lineages. Questions about "how human" an organism must be before meriting the treatment a human would deserve have also come under discussion. These and other issues require careful and comprehensive multidisciplinary discussions to evaluate risks, include safety measures, and protect against potential misuse.

6. CONCLUDING REMARKS

CRISPR–Cas9 technology has accelerated and enabled the production of a wide range of animals, and this has led to discoveries and innovations throughout basic and applied science. CRISPR–Cas9 technology has not only contributed to expand the number of animals that are tractable for genome modification, but has also enabled the generation of highly sophisticated genotypes. In view of the impressive achievements made in the last few years with this technology, we expect to see many novel discoveries and products involving genetically modified animals in the next years.

REFERENCES

1. Rocha-Martins M, Cavalheiro GR, Matos-Rodrigues GE, Martins RAP. From gene targeting to genome editing: transgenic animals applications and beyond. *Acad Bras Cienc.* 2015;87(2):1323–1348.
2. Doudna JA, Charpentier E. Genome editing. The new frontier of genome engineering with CRISPR-Cas9. *Science.* 2014;346(6213):1258096.
3. Mali P, Esvelt KM, Church GM. Cas9 as a versatile tool for engineering biology. *Nat Methods.* 2013;10(10):957–963.
4. Hsu PD, Lander ES, Zhang F. Development and applications of CRISPR-Cas9 for genome engineering. *Cell.* 2014;157(6):1262–1278.
5. Sander JD, Joung JK. CRISPR-Cas systems for editing, regulating and targeting genomes. *Nat Biotechnol.* 2014;32(4):347–355.
6. Barrangou R, Fremaux C, Deveau H, et al. CRISPR provides acquired resistance against viruses in prokaryotes. *Science.* 2007;315(5819):1709–1712.
7. Wiedenheft B, Sternberg SH, Doudna JA. RNA-guided genetic silencing systems in bacteria and archaea. *Nature.* 2012;482(7385):331–338.
8. Jinek M, East A, Cheng A, Lin S, Ma E, Doudna J. RNA-programmed genome editing in human cells. *Elife.* 2013;2:e00471.
9. Mali P, Yang L, Esvelt KM, et al. RNA-guided human genome engineering via Cas9. *Science.* 2013;339(6121):823–826.
10. Cong L, Ran FA, Cox D, et al. Multiplex genome engineering using CRISPR/Cas systems. *Science.* 2013;339(6121):819–823.
11. Chu VT, Weber T, Wefers B, et al. Increasing the efficiency of homology-directed repair for CRISPR-Cas9-induced precise gene editing in mammalian cells. *Nat Biotechnol.* 2015;33(5):543–548.
12. Zhang J-P, Li X-L, Li G-H, et al. Efficient precise knockin with a double cut HDR donor after CRISPR/Cas9-mediated double-stranded DNA cleavage. *Genome Biol.* 2017;18(1):35.
13. Davis L, Maizels N. Homology-directed repair of DNA nicks via pathways distinct from canonical double-strand break repair. *Proc Natl Acad Sci USA.* 2014;111(10):E924–E932.

14. Byrne SM, Ortiz L, Mali P, Aach J, Church GM. Multi-kilobase homozygous targeted gene replacement in human induced pluripotent stem cells. *Nucl Acids Res.* 2014;43(3):e21.
15. Fu Y, Foden JA, Khayter C, et al. High-frequency off-target mutagenesis induced by CRISPR-Cas nucleases in human cells. *Nat Biotechnol.* 2013;31(9):822–826.
16. Lin Y, Cradick TJ, Brown MT, et al. CRISPR/Cas9 systems have off-target activity with insertions or deletions between target DNA and guide RNA sequences. *Nucl Acids Res.* 2014;42(11):7473–7485.
17. Frock RL, Hu J, Meyers RM, Ho Y-J, Kii E, Alt FW. Genome-wide detection of DNA double-stranded breaks induced by engineered nucleases. *Nat Biotechnol.* 2015;33 (2):179–186.
18. Aach J, Church GM, Mali P, Church GM. *CasFinder: Flexible Algorithm for Identifying Specific Cas9 Targets in Genomes.* New York, NY: Cold Spring Harbor Labs Journals; 2014.
19. Schumann K, Lin S, Boyer E, et al. Generation of knock-in primary human T cells using Cas9 ribonucleoproteins. *Proc Natl Acad Sci.* 2015;112(33):201512503.
20. Fu Y, Sander JD, Reyon D, Cascio VM, Joung JK. Improving CRISPR-Cas nuclease specificity using truncated guide RNAs. *Nat Biotechnol.* 2014;32(3):279–284.
21. Kleinstiver BP, Pattanayak V, Prew MS, et al. High-fidelity CRISPR-Cas9 nucleases with no detectable genome-wide off-target effects. *Nature.* 2016;529(7587): 490–495.
22. Slaymaker IM, Gao L, Zetsche B, Scott DA, Yan WX, Zhang F. Rationally engineered Cas9 nucleases with improved specificity. *Science.* 2015;351(6268):84–88.
23. Ran FA, Hsu PD, Lin C-Y, et al. Double nicking by RNA-guided CRISPR Cas9 for enhanced genome editing specificity. *Cell.* 2013;154(6):1380–1389.
24. Guilinger JP, Thompson DB, Liu DR. Fusion of catalytically inactive Cas9 to FokI nuclease improves the specificity of genome modification. *Nat Biotechnol.* 2014;32 (6):577–582.
25. Ran FA, Cong L, Yan WX, et al. In vivo genome editing using *Staphylococcus aureus* Cas9. *Nature.* 2015;520(7546):186–191.
26. Kato T, Takada S. In vivo and in vitro disease modeling with CRISPR/Cas9. *Brief Funct Genomics.* 2017;16:13–24.
27. Friedland AE, Tzur YB, Esvelt KB, Colaiácovo MP, Church GM, Calarco JA. Heritable genome editing in *C. elegans* via a CRISPR-Cas9 System. *Nat Methods.* 2013;10: 741–743.
28. Yu Z, Ren M, Wang Z, et al. Highly efficient genome modifications mediated by CRISPR/Cas9 in *Drosophila. Genetics.* 2013;195(1):289–291.
29. Hammond A, Galizi R, Kyrou K, et al. A CRISPR-Cas9 gene drive system targeting female reproduction in the malaria mosquito vector *Anopheles gambiae. Nat Biotechnol.* 2015;34(1):78–83.
30. Lin C-Y, Su Y-H. Genome editing in sea urchin embryos by using a CRISPR/Cas9 system. *Dev Biol.* 2016;409(2):420–428.
31. Li M, Zhao L, Page-McCaw PS, Chen W. Zebrafish genome engineering using the CRISPR–Cas9 system. *Trends Genet.* 2016;32(12):815–827.
32. Ikmi A, McKinney SA, Delventhal KM, Gibson MC, Stuurman N. TALEN and CRISPR/Cas9-mediated genome editing in the early-branching metazoan *Nematostella vectensis. Nat Commun.* 2014;5:5486.
33. Wang H, Yang H, Shivalila CS, et al. One-step generation of mice carrying mutations in multiple genes by CRISPR/Cas-mediated genome engineering. *Cell.* 2013;153 (4):910–918.
34. Niu Y, Shen B, Cui Y, et al. Generation of gene-modified cynomolgus monkey via Cas9/RNA-mediated gene targeting in one-cell embryos. *Cell.* 2014;156:836–843.

35. Ma H, Nuria M-G, Sang-Wook P, Jun W, Yeonmi L, Keiichiro S, Amy K, et al. Correction of a Pathogenic Gene Mutation in Human Embryos. *Nature.* 2017.
36. Dickinson DJ, Goldstein B. CRISPR-based methods for *Caenorhabditis elegans* genome engineering. *Genetics.* 2016;202(3):885–901.
37. Hwang WY, Fu Y, Reyon D, et al. Efficient genome editing in zebrafish using a CRISPR-Cas system. *Nat Biotechnol.* 2013;31(3):227–229.
38. Varshney GK, Pei W, Lafave MC, et al. High-throughput gene targeting and phenotyping in zebrafish using CRISPR/Cas9. *Genome Res.* 2015;25:1030–1042.
39. Sánchez-Rivera FJ, Papagiannakopoulos T, Romero R, et al. Rapid modelling of cooperating genetic events in cancer through somatic genome editing. *Nature.* 2014;516 (7531):428–431.
40. Swiech L, Heidenreich M, Banerjee A, et al. In vivo interrogation of gene function in the mammalian brain using CRISPR-Cas9. *Nat Biotechnol.* 2014;33(1):102–106.
41. Cheng R, Peng J, Yan Y, et al. Efficient gene editing in adult mouse livers via adenoviral delivery of CRISPR/Cas9. *FEBS Lett.* 2014;588(21):3954–3958.
42. Zuris JA, Thompson DB, Shu Y, et al. Cationic lipid-mediated delivery of proteins enables efficient protein-based genome editing in vitro and in vivo. *Nat Biotechnol.* 2014;33(1):73–80.
43. Li P, Estrada JL, Burlak C, et al. Efficient generation of genetically distinct pigs in a single pregnancy using multiplexed single-guide RNA and carbohydrate selection. *Xenotransplantation.* 2015;22:20–31.
44. Gao Y, Wu H, Wang Y, et al. Single Cas9 nickase induced generation of NRAMP1 knockin cattle with reduced off-target effects. *Genome Biol.* 2017;18(1):13.
45. Verma G, Arora JS, Sethi RS, Mukhopadhyay CS, Verma R. Handmade cloning: recent advances, potential and pitfalls. *J Anim Sci Biotechnol.* 2015;6:43.
46. Wang Y, Du Y, Zhou X, et al. Efficient generation of B2m-null pigs via injection of zygote with TALENs. *Sci Rep.* 2016;6(1):38854.
47. Chan AWS. Progress and prospects for genetic modification of nonhuman primate models in biomedical research. *ILAR J.* 2013;54(2):211–223.
48. Chan AW, Chong KY, Martinovich C, Simerly C, Schatten G. Transgenic monkeys produced by retroviral gene transfer into mature oocytes. *Science.* 2001;291(5502):309–312.
49. Niu Y, Yu Y, Bernat A, et al. Transgenic rhesus monkeys produced by gene transfer into early-cleavage-stage embryos using a simian immunodeficiency virus-based vector. *Proc Natl Acad Sci USA.* 2010;107(41):17663–17667.
50. Sasaki E, Suemizu H, Shimada A, et al. Generation of transgenic non-human primates with germline transmission. *Nature.* 2009;459(7246):523–527.
51. Shalem O, Sanjana NE, Hartenian E, et al. Genome-scale CRISPR-Cas9 knockout screening in human cells. *Science.* 2014;343(6166):84–87.
52. Zhou Y, Zhu S, Cai C, et al. High-throughput screening of a CRISPR/Cas9 library for functional genomics in human cells. *Nature.* 2014;509(7501):487–491.
53. Shah AN, Davey CF, Whitebirch AC, Miller AC, Moens CB. Rapid reverse genetic screening using CRISPR in zebrafish. *Nat Methods.* 2015;12(6):535–540.
54. D'Agostino Y, Locascio A, Ristoratore F, et al. A rapid and cheap methodology for CRISPR/Cas9 zebrafish mutant screening. *Mol Biotechnol.* 2016;58(1):73–78.
55. Yang L, Guell M, Byrne S, et al. Optimization of scarless human stem cell genome editing. *Nucl Acids Res.* 2013;41(19):9049–9061.
56. Manguso RT, Pope HW, Zimmer MD, Brown FD, Yates KB, Miller BC, Collins NB, et al. In Vivo CRISPR Screening Identifies Ptpn2 as a Cancer Immunotherapy Target. *Nature.* 2017;547(7664):413–418.
57. Canver MC, Bauer DE, Dass A, et al. Characterization of genomic deletion efficiency mediated by clustered regularly interspaced palindromic repeats (CRISPR)/Cas9 nuclease system in mammalian cells. *J Biol Chem.* 2014;289(31):21312–21324.

58. Wang L, Shao Y, Guan Y, et al. Large genomic fragment deletion and functional gene cassette knock-in via Cas9 protein mediated genome editing in one-cell rodent embryos. *Sci Rep.* 2015;5:17517.
59. Seruggia D, Fernández A, Cantero M, Pelczar P, Montoliu L. Functional validation of mouse tyrosinase non-coding regulatory DNA elements by CRISPR-Cas9-mediated mutagenesis. *Nucl Acids Res.* 2015;43(10):4855–4867.
60. Dow LE. Modeling disease in vivo with CRISPR/Cas9. *Trends Mol Med.* 2015;21 (10):609–621.
61. Mei Y, Wang Y, Chen H, Sun ZS, Ju X-D. Recent progress in CRISPR/Cas9 technology. *J Genet Genom.* 2016;43(2):63–75.
62. Platt RJ, Chen S, Zhou Y, et al. CRISPR-Cas9 knockin mice for genome editing and cancer modeling. *Cell.* 2014;159(2):440–455.
63. Xue W, Chen S, Yin H, et al. CRISPR-mediated direct mutation of cancer genes in the mouse liver. *Nature.* 2014;514(7522):380–384.
64. Maddalo D, Manchado E, Concepcion CP, et al. In vivo engineering of oncogenic chromosomal rearrangements with the CRISPR/Cas9 system. *Nature.* 2014;516 (7531):423–427.
65. Heckl D, Kowalczyk MS, Yudovich D, et al. Generation of mouse models of myeloid malignancy with combinatorial genetic lesions using CRISPR-Cas9 genome editing. *Nat Biotechnol.* 2014;32(9):941–946.
66. Tabebordbar M, Zhu K, Cheng JKW, et al. In vivo gene editing in dystrophic mouse muscle and muscle stem cells. *Science.* 2016;351(6271):407–411.
67. Nelson CE, Hakim CH, Ousterout DG, et al. In vivo genome editing improves muscle function in a mouse model of Duchenne muscular dystrophy. *Science.* 2016;351 (6271):403–407.
68. Ousterout DG, Kabadi AM, Thakore PI, Majoros WH, Reddy TE, Gersbach CA. Multiplex CRISPR/Cas9-based genome editing for correction of dystrophin mutations that cause Duchenne muscular dystrophy. *Nat Commun.* 2015;6:6244.
69. Long C, McAnally JR, Shelton JM, Mireault AA, Bassel-Duby R, Olson EN. Prevention of muscular dystrophy in mice by CRISPR/Cas9-mediated editing of germline DNA. *Science.* 2014;345(6201):1184–1188.
70. Zhu P, Wu F, Mosenson J, Zhang H, He T-C, Wu W-S. CRISPR/Cas9-mediated genome editing corrects dystrophin mutation in skeletal muscle stem cells in a mouse model of muscle dystrophy. *Mol Ther Nucl Acids.* 2017;7:31–41.
71. Patel M. Cows turned white for meat that beats the heat News The Times & The Sunday Times. http://www.aggenetics.com/ag2/wp-content/uploads/2016/05/Cows-turned-white-for-meat-that-beats-the-heat-_-News-_-The-Times-The-Sunday-Times.pdf; 2017
72. Carlson DF, Lancto CA, Zang B, et al. Production of hornless dairy cattle from genome-edited cell lines. *Nat Biotechnol.* 2016;34(5):479–481.
73. Dhanapala P, Doran T, Tang MLK, Suphioglu C. Production and immunological analysis of IgE reactive recombinant egg white allergens expressed in *Escherichia coli.* *Mol Immunol.* 2015;65(1):104–112.
74. Oishi I, Yoshii K, Miyahara D, Kagami H, Tagami T. Targeted mutagenesis in chicken using CRISPR/Cas9 system. *Sci Rep.* 2016;6:23980.
75. Burkard C, Lillico SG, Reid E, et al. Precision engineering for PRRSV resistance in pigs: Macrophages from genome edited pigs lacking CD163 SRCR5 domain are fully resistant to both PRRSV genotypes while maintaining biological function. *PLoS Pathog.* 2017;13(2):e1006206. [Perez DR, ed.].
76. Ma X, Zhu Q, Chen Y, Liu Y-G. CRISPR/Cas9 platforms for genome editing in plants: developments and applications. *Mol Plant.* 2016;9(7):961–974.

77. Monzani PS, Adona PR, Ohashi OM, Meirelles FV, Wheeler MB. Transgenic bovine as bioreactors: challenges and perspectives. *Bioengineered.* 2016;7(3):123–131.
78. Cyranoski D. Gene-edited "micropigs" to be sold as pets at Chinese institute. *Nature.* 2015;526(7571):18.
79. Reardon S. Welcome to the CRISPR zoo. *Nature.* 2016;531(7593):160–163.
80. Zou Q, Wang X, Liu Y, et al. Generation of gene-target dogs using CRISPR/Cas9 system. *J Mol Cell Biol.* 2015;7(6):580–583.
81. Shafran D, Kodish E, Tzakis A. Organ shortage: the greatest challenge facing transplant medicine. *World J Surg.* 2014;38(7):1650–1657.
82. Vadori M, Cozzi E. The immunological barriers to xenotransplantation. *Tissue Antigens.* 2015;86(4):239–253.
83. Semaan M, Ivanusic D, Denner J. Cytotoxic effects during knock out of multiple porcine endogenous retrovirus (PERV) sequences in the pig genome by zinc finger nucleases (ZFN). *PLoS One.* 2015;10(4):e0122059.
84. Cooper DK, Ekser B, Ramsoondar J, Phelps C, Ayares D. The role of genetically engineered pigs in xenotransplantation research. *J Pathol.* 2016;238(2):288–299.
85. Whitworth KM, Lee K, Benne JA, et al. Use of the CRISPR/Cas9 system to produce genetically engineered pigs from in vitro-derived oocytes and embryos. *Biol Reprod.* 2014;91(3):78.
86. Yang L, Güell M, Niu D, et al. Genome-wide inactivation of porcine endogenous retroviruses (PERVs). *Science.* 2015;350(6264).
87. Niu D, Hong-Jiang W, Lin L, Haydy G, Tao W, I-Hsiu L, Hong-Ye Z, et al. Inactivation of porcine endogenous retrovirus in pigs using CRISPR-Cas9. *Science.* 2017.
88. Iwase H, Liu H, Wijkstrom M, et al. Pig kidney graft survival in a baboon for 136 days: longest life-supporting organ graft survival to date. *Xenotransplantation.* 2015;22 (4):302–309.
89. Kobayashi T, Yamaguchi T, Hamanaka S, et al. Generation of rat pancreas in mouse by interspecific blastocyst injection of pluripotent stem cells. *Cell.* 2010;142(5):787–799.
90. Jun Wu A, Platero-Luengo A, Sakurai M, et al. Interspecies chimerism with mammalian pluripotent stem cells. *Cell.* 2017;168:473–486.
91. Piaggio AJ, Segelbacher G, Seddon PJ, et al. Is it time for synthetic biodiversity conservation? *Trends Ecol Evol.* 2017;32(2):97–107.
92. Simoni A, Siniscalchi C, Chan YS, et al. Development of synthetic selfish elements based on modular nucleases in *Drosophila melanogaster. Nucl Acids Res.* 2015;43(5):2991.
93. Bassett AR, Tibbit C, Ponting CP, Liu J-L. Highly efficient targeted mutagenesis of *Drosophila* with the CRISPR/Cas9 system. *Cell Rep.* 2013;4(1):220–228.
94. Gratz SJ, Cummings AM, Nguyen JN, et al. Genome engineering of *Drosophila* with the CRISPR RNA-guided Cas9 nuclease. *Genetics.* 2013;194(4):1029–1035.
95. Gantz VM, Bier E. Genome editing. The mutagenic chain reaction: a method for converting heterozygous to homozygous mutations. *Science.* 2015;348(6233):442–444.
96. Hammond A, Galizi R, Kyrou K, et al. A CRISPR-Cas9 gene drive system targeting female reproduction in the malaria mosquito vector *Anopheles gambiae. Nat Biotechnol.* 2016;34(1):78–83.
97. Gantz VM, Jasinskiene N, Tatarenkova O, et al. Highly efficient Cas9-mediated gene drive for population modification of the malaria vector mosquito *Anopheles stephensi. Proc Natl Acad Sci USA.* 2015;112(49):E6736–E6743.
98. Capizzi D, Bertolino S, Mortelliti A. Rating the rat: global patterns and research priorities in impacts and management of rodent pests. *Mamm Rev.* 2014;44(2):148–162.
99. Unckless RL, Clark AG, Messer PW. *Evolution of resistance against CRISPR/Cas9 gene drive.* bioRxiv. Available from: http://biorxiv.org/content/early/2016/06/11/058438; 2016
100. Shultz D. Should we bring extinct species back from the dead? *Science*; September 2016

101. Shapiro B. Mammoth 2.0: will genome engineering resurrect extinct species? *Genome Biol.* 2015;16(1):228.
102. Folch J, Cocero MJ, Chesné P, et al. First birth of an animal from an extinct subspecies (*Capra pyrenaica pyrenaica*) by cloning. *Theriogenology.* 2009;71(6):1026–1034.
103. Lynch VJ, Bedoya-Reina OC, Ratan A, et al. Elephantid genomes reveal the molecular bases of woolly mammoth adaptations to the Arctic. *Cell Rep.* 2015;12(2):217–228.
104. Church GM, Regis E. *Regenesis: How Synthetic Biology Will Reinvent Nature and Ourselves.* New York, NY: Basic Books; 2012. https://books.google.com/books/about/Regenesis.html?id=sWahIAKG604C
105. Combes RD, Balls M. Every silver lining has a cloud: the scientific and animal welfare issues surrounding a new approach to the production of transgenic animals. *Altern Lab Anim.* 2014;42(2):137–145. http://www.ncbi.nlm.nih.gov/pubmed/24901907
106. Rodriguez E. Ethical issues in genome editing using Crispr/Cas9 system. *J Clin Res Bioeth.* 2016;7(2):1–4.

CHAPTER SEVEN

Gene Editing and CRISPR Therapeutics: Strategies Taught by Cell and Gene Therapy

Juan C. Ramirez[1]
VIVEbioTECH, San Sebastián, Spain
[1] Corresponding author. E-mail address: jcramirez@vivebiotech.com

Contents

Abstract

A few years ago, we assisted in the demonstration for the first time of the revolutionary idea of a type of adaptive-immune system in the bacteria kingdom. This system, named CRISPR, and variants engineered in the lab, have been demonstrated as functional with extremely high frequency and fidelity in almost all eukaryotic cells studied to date. The capabilities of this RNA-guided nuclease have added to the interest that was announced with the advent of previous technologies for genome editing tools, such as ZFN and TALEN. The capabilities exhibited by these gene editors, opens up a novel scenario that indicates the promise of a next-generation medicine based on precision and personalized objectives, mostly due to the change in the paradigm regarding gene-surgery. This has certainly attracted, like never before, the attention of the biotech business and investor community. This chapter offers a brief overview of some of the factors that have contributed to a rapid entry into the biotech and pharmaceutical company's pipeline, focusing on how cell and gene therapies (CGT), collectively known as advanced therapies, have become the driving forces toward the therapeutic uses of gene editing technology. The sum of all those efforts for more than 30 years has contributed to the new paradigm of considering genes as medicines.

Progress in Molecular Biology and Translational Science, Volume 152
ISSN 1877-1173
http://dx.doi.org/10.1016/bs.pmbts.2017.08.003

1. GENOME EDITORS ENTER THE SCENE

Targeted genome modification using designer nucleases is an emerging technology that can be used to investigate gene function and could also be used to treat genetic or acquired diseases.[1] Recently developed genomic editing technologies, such as zinc finger nucleases (ZFNs), transcription activator-like effector nucleases (TALENs), meganucleases, and clustered regularly interspaced short palindromic repeats (CRISPR) have the potential to be powerful tools in gene therapy because of their ability to inactivate genes, correct mutated sequences, or insert intact genes.[2]

Indeed, ZFNs are the first class of nucleases to have reached the clinic in phase I trials for HIV[3], and other recently developed nucleases are close to reaching the clinic.[4] One key issue, that will be addressed as these technologies, move into clinical trials, whether technology-specific preclinical evaluations are available, that can establish safety regarding their precision. Specifically, the degree to which, there are off-target actions and the clinical implications of such activity are important questions for the field.

Other chapters of this book have presented excellent reviews regarding the state of the art of genome-editing technologies, and in particular regarding RNA-guided engineered nucleases (RGEN), such as CRISPR, providing an analysis from the technical point of view, weighing up the pros and cons of their uses in research, and the differences from other available alternative gene editing approaches. In this chapter, the author addresses some of the factors that have contributed to an unprecedented rapid entry into the biotech and pharmaceutical company's pipeline for advanced therapy development, and which are consequently becoming the driving forces toward the therapeutic uses of gene editing technology. The analysis focuses on the unique, outstanding features of the development of the gene-therapy field in relation to biotech companies, the safety requirements addressed both before and upon their entry into clinics, and relationships with the evolution of safe and efficient delivery systems for gene-based products and the interaction between the academy and industry as outstanding players in the current scenario regarding the advent of gene editing. Genome editing capabilities have captured the attention of the popular media, scientific community, and the investment world: the promise of going into a cell and making precise changes to their genetic code to cure them of illnesses is certainly captivating and, beyond that, raising ethical concerns.[5] The opportunities to treat disease and improve health are indeed tremendous.

This technology is without doubt, one of the major molecular biology revolutions. And it is cheap and easy.

However, beyond being a revolutionary research tool, CRISPR also has enormous potential for changing medicine, and this has certainly attracted the attention of the biotech business and investor community.

This review intends to present, at a glance, several perspectives for gene-editing technologies entering biotech, and awake the interest not only of small companies but also of big pharma. Despite the fact that the gene editing approach has been shown to be useful and has been applied successfully to more than 50 different organisms, including crop plants, livestock, and humans, this review will focus only on certain aspects of the biotech industry which is engaged in gene editing for human health. The review is divided into three short sections: first, to adjust the gene editing technique to Gartner hype–hope curve for technologies, and see how it compares to other previously developed advanced therapies; second, to sketch the manner in which industry and academia have been developing cell and gene therapy technologies, and how they have assisted in enabling genome-editing technologies to reach businesses faster than ever; and third, a personal summary of the challenges that biotech and pharma industries, health-care services and society will face with the advent of such powerful and revolutionary gene editing technologies aimed at changing our perception of medicine and the world we live in.

2. MAKING ADVANCED TECHNOLOGIES GROUND-BREAKING

Since the completion of the Human Genome Project in 2003, and the publication of a reference human genome sequence, genomics has become a mainstay of biomedical research. Optimism about the potential contributions of genomics toward improving human health has been fueled by new insights about diseases, the molecular basis of inherited diseases, and the role of structural variations in diseases, some of which have already led to new therapies. Other advances have already changed medical practice. For example, microarrays are now used for the clinical detection of genomic imbalances, and pharmacogenomic testing is routinely performed before the administration of certain medications. Together, these achievements provide evidence that genomics is contributing toward a better understanding of human biology and an improvement in human health.

Cell and gene therapies (CGT) collectively known as advanced therapies, since their origin in the very late 80's, changed the paradigm by considering genes, not only as the ultimate basis for the disease, but also as the ultimate solution in repairing cells and curing many diseases, coining the idea of genes as medicines. The advent of these capabilities in gene editing opens up a novel scenario that points toward the promise of a next generation medicine based on precision and personalized objectives, mainly due to the change in the paradigm regarding gene-surgery.

Despite CGTs being pioneering in such a change to the medicine paradigm, they have also changed the relationship between the biotech/pharma industry and academia, regarding progress in research and commercialization of health products. CGTs represent an outside–in technology[6], as opposed to the conventional business practices that have led to the creation of traditional medicines that traveled from the inside of the exporting industry, and drove the innovation in the productive and discovery strategies toward academia.

These small molecule and protein-based drugs were largely developed within the BioPharma industry. While key discoveries often came from academia (genetic engineering technology, recombinant proteins, monoclonal antibodies, and phage display), their route into clinical practice has historically been the sphere of industry. The know-how and skills to go from basic discovery to formal "early development" and clinical medicine mostly emerged from industry-driven efforts. New chemical entities were the products of fully integrated research, development, and manufacturing teams. All the expertise were largely owned and developed in-house, and the art of biomedical product promotion and bioprocessing was moved forward through in-house learning within the biotechnology industry. The story has lasted over time and academic discoveries, such as phage display and transgenic mice were introduced to the industry in their developmental stages and included in their industrial discovery and early development programs, through clinical studies, and eventually to the market. Insights from the industrial transformation process were adopted by academia, which in the USA is exemplified by the Scripps and the Broad Institutes that recruited exindustry drug discovery to provided their know-how, their strategies and discovery models, in order to reach objectives.[6] In the end, academia benefited from industry insights and incorporated their development schemes designed for drug discovery, together with their basic research programs.

The CGT, in contrast, is an outside–in model of technological development. Key advances in gene and cell therapy and most of the breakthroughs in the field were made by pioneer researchers in cell and gene therapy working in academia, which directed the development of translational models of their discoveries toward the market, and they deserve to be recognized as real transformation leaders. Only after their seminal publications on efficacy and safety attracted the attention of companies like Novartis, Pfizer, or Bayer, did significant biotech and venture capital funding allow the launch of early- and late-clinical gene therapy companies, such as Juno Therapeutics, Spark Therapeutics, Kite Pharma, AGCT, RegenXBio, Asklepios BioPharmaceuticals, and many others. Remarkably, the repeated circle of research, and preclinical models was driven again and again by those pioneer academics, until they demonstrated clinical efficacy as a necessary step in marketing their ground-breaking therapies.

In contrast to the conventional drug development process that seeded the mainstream CRO/CMO industry products prior to the industry insight, many of the CGT players do not have in-house manufacturing space for viruses or stem cells, and these necessary steps rely on multiple vendors which exist externally, representing another differentiating feature of the innovative outside-in model. Finally, the synergy and convergence of technologically independent fields has been one of the most important factors contributing toward accelerating the advance of advanced therapies. Let's see how it happened.

3. HOW HAVE WE REACHED THIS POINT SO QUICKLY?

Gene-editing technologies are generating huge expectations for human health, because they are presumably expected to quickly appear in the clinic: the media, society, therefore, as a consequence, venture funding, has put the focus on the technology, and its uses. It is thus worth considering, how other advanced therapies, based on CGT have reach the market after almost 3 decades of effort. Are the predictions accurate, and which factors are contributing toward accelerating the process?

The Gartner Hype Cycle (GH cycle) (Fig. 1) has been applied in both the technological and biotechnology sectors.[7] The curve is a predictive representation of the behavior of the endpoint for a new technology. Most advanced technologies follow the shape of the GH cycle, with exponential growth during the early stages, fed by high expectations, and a subsequent downward profile caused by unexpected situations. The mature plateau-shaped stage is preceded by a slope representing the introduction of improvements and the reevaluation of critical factors that were the basis

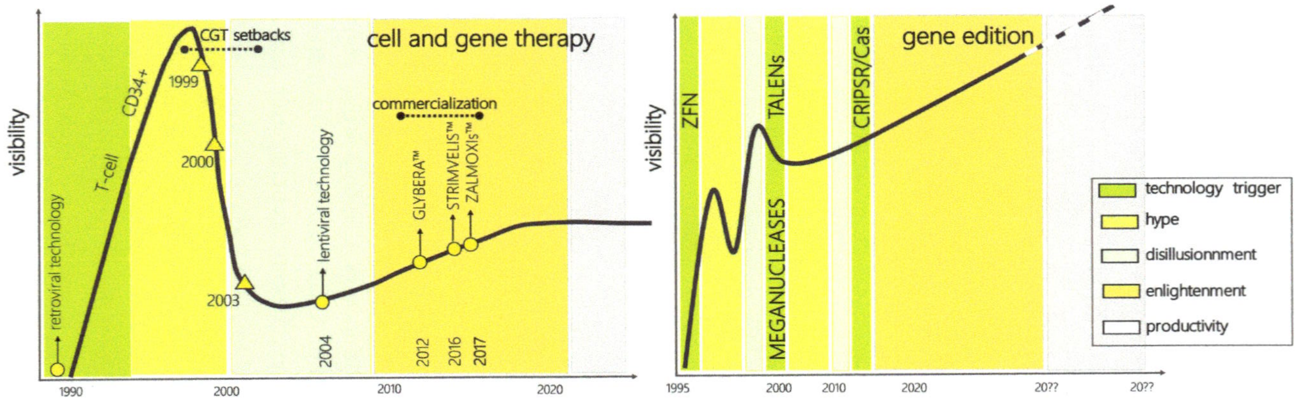

Fig. 1 The Gartner Hype–Hope cycle is represented for cell and gene therapy (left) and gene editing technologies (right). The color code represents the five phases into which Gartner's curves are divided. Milestones at several phases during the development of CGT are indicated, and are referenced in the text. The gene editing technology curve tries to reflect the superposition and overlapping of different technologies that appeared over almost a decade (2000–12), and to comment on how fast the technology is expected to reach the final phases (slope of enlightenment). This trend is supported by the IPO for CRISPR-based companies less than 2 years after they were found (Fig. 3).

for the preceding fall. CGT events occurring since 1990 fit such a cycle (Fig. 1; left panel). In the early 90's ADA-SCID (a fatal metabolic disorder affecting the immune system, such as the bubble-boy disease) was considered the ideal target for gene therapy because the genetic basis of the disease was understood; the disease could be corrected by gene modification in preclinical models; and production of normal protein at low levels from corrected cells conferred a survival advantage. The knowledge in T-cell biology and the model disease prompted the NIH in 1990 to launch the first gene therapy trial. Failure came from the low maintenance of the corrected cells. The peak in the GH cycle was reached with the work on CD34+ stem cell gene therapy and clinical trials carried out by scientists at the San Raffaele Hospital, Milan, starting in 1992, which increased expectations until the year 2000. These events prompted the development of more academic trials in more diseases, but unexpected deaths due to the treatment of two pediatric patients in France with a first-generation retroviral vector for a rare immunosuppression (X-SCID)[8] and an adult patient from a rare hepatic disorder in USA treated with an adenoviral vector[9] shocked the gene therapy field, demonstrating that the approach was not ready (slope of disillusionment). This was the turning point for CGT, with huge consequences for the field, including a deep change in the development model and significantly, the field was relegated to the periphery of industrial biomedical research. In such a scenario, VC's were not very keen on funding these companies, and pharma gave up the business model of selling cells and viruses as products. As described in the previous section, these issues were supported by the outside–in model of development. Researchers retreated to their laboratories to learn from those failures, and to come up with more safe vectors, stronger clinical safety protocols, and stronger and more accurate supporting data from both the disease and the animal models. It took over a decade to climb the slope of the hope curve.

The period 2005–10 was the slope of progress for CGT (Fig. 1; left panel). Three milestones deserve a special mention[10,11]: (1) data from the ongoing ADA-SCID trial at San Raffaele Hospital, (2) positive results from patients with adrenoleukodystrophy and beta thalassemia treated with gene therapy, and (3) the take-off of biotech companies embarked on the development of gene therapy programs, such as in Oxford, the UK-based Oxford BioMedica and in Cambridge, MA, the USA-based Bluebird Bio, which were found in the 90's (bluebird was created as Genetix) as companies that worked out-of-mainstream when industrial R&D neglected advances in the field. New results arose from the use of lentiviral vectors that demonstrated safety profiles and became the gold-standard of stable vectors for dividing CD34+ cells gene therapy.[12] The plateau in the GH cycle was sustained by the entry of big pharma GSK as partnership with San Raffaele/TIGET (Milan, Italy) to carry out the final-stage development of ADA-SCID, and the establishment of a gene therapy platform based on lentiviral/stem CD34 + gene therapy with multiple programs. After over 25 years of research, a new type of medicine was approved in Europe at the end of May 2016, resulting from the successful convergence of two fields of science: stem cell transplantation and gene therapy. This is Strimvelis, a GSK patient-specific gene-modified stem cell medicine for ADA-SCID.[13]

From the technological point of view, the current CGT scenario on the enlightenment slope of the Gardner's cycle is due to a synergy of efforts and advances in several areas. After more than 50 years of transplant medicine, millions of patients have received transplants globally. This medical advance has boosted knowledge regarding ex vivo cell manipulation of stem cells and/or T-cells, synergized with an impressive advance in viral vectorology and nucleic acid-delivery technologies, and promoted a renaissance in gene therapy, and the maturation of the field.

The clinical experience of CGT is of major value to efficient translational studies of gene editing approaches. In addition, nowadays the engine driving the next phase of technological innovation is at least in part the engagement of early-biotech owning innovative developments and technologies in the field, and their partnerships with big companies that will help to promptly transform these developments into products (Fig. 2). However, all these new technologies must still face pending challenges, such as cost-efficient manufacturing strategies, or overcome the barriers of expensive infrastructure and know-how

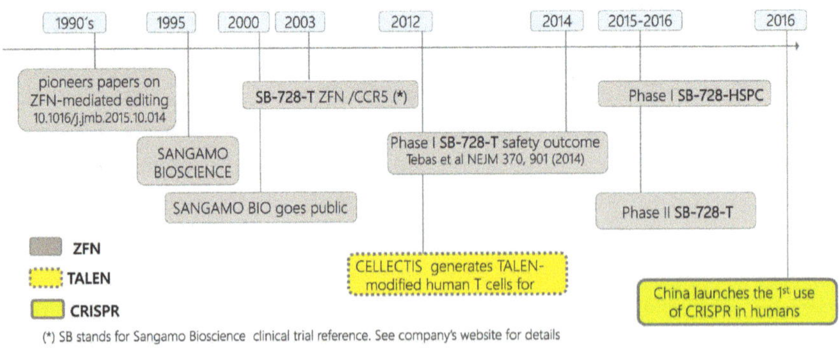

Fig. 2 Milestones in therapeutic uses of gene editing technologies.

that contribute to the elevated cost of treatments, a central issue in the actual transfer to market of CGT and gene editing technologies. New science will be needed to transform the field in the coming years, and success will partly depend on partnering capable of create strategies aimed at entering potentially geographical, disease-related, or technological goals. Our expectancy is to witness similar synergies between viral gene therapy and cell transplant therapy, with gene editing to constitute a new business model capable of driving a virtuous cycle to promote these new innovative therapies.

4. HOW FAR INTO THE CLINIC WILL GENOME-EDITING GO?

Gene editing is a fascinating the medicine of the future, opening a window to actual personalized precision medicine. Genome engineering is becoming the scalpel, which is capable of reaching the cause of a genetic disease; a scenario resembling that of surgeons repairing damaged tissues and organs, but far removed from the nonsurgical interventions regularly used to treat the symptoms and the side-effects of the disease. The point is, how to transform the basic research results into a commercially valuable product. Let us briefly explore, in the next section, what has been done and what are some of the risks that these businesses are facing.

The CRISPR/Cas technique, as with the previous ZFN and TALEN techniques, has matured tremendously in less than a decade and has demonstrated its feasibility as a widely applicable technique in developmental programs and research, demonstrating robustness and relative safety behavior in these contexts (Fig. 2). It is outside the scope of this review to go deeper into these issues. However, it is necessary to mention that there are still core-open questions that must be answered before it becomes a clinical technology, involving patients on trials, beyond the palliative interventions recently approved in China.[14] Three aspects are important to fully develop and propel the potential of genome editing as a medicinal product: (1) an efficient, safe, hit and run delivery of gene editing technologies, (2) specificity of gene editing technologies, and (3) efficiency of gene editing both ex vivo and in vivo (Table 1). The hallmark of the nuclease of the future capable of propelling the field forward must provide high-specificity and affinity for its cognate binding site such that the mass action equilibrium will not shift in favor of off-target sites with small increases of nuclease concentration, in contrast to bad nucleases with low specificity that are prone to bind more

Table 1 Major Aspects in the CRISPR/Cas System That Require Improvements

Feature	Needs
Nuclease engineering	Identify and optimize both Cas9 enzymes and guide RNA molecules, to create the optimal Cas9-guide RNA complex for a given disease target
Delivery	An appropriate product configuration must be designed to provide efficient and tightly controlled delivery to the desired tissue or cell type
Control and specificity	Control of cellular exposure to the Cas9-guide RNA complex and specificity of the DNA cut are important to optimizing the location and duration of editing activity
Directed editing	Different mechanisms that a cell uses to repair cuts in DNA results in different kinds of genetic changes. Approaches to selectively harnessing specific DNA repair mechanisms to be able to drive the appropriate type of repair for a given disease

off-target sites with small changes in nuclease concentration, limiting the ability to reach an effective nontoxic dose.[15] Unfortunately, this good nuclease is still to be discovered, or more likely engineered, despite the fact that new formats are being discovered and modified monthly.[16]

Regardless of future developments in the RNA-guided nuclease systems, gene editing also has a long history, originating in zinc-finger modification of the genome in 1996[1] (Fig. 2). Sangamo Therapeutics has been a pioneer with this technology and demonstrated broad success at selective modification of multiple genes using this technology as a research tool. Remarkably, Sangamo has reported proof-of-concept clinical studies to treat HIV infection using an engineered ZFN to inactivate the gene coding for the CC chemokine receptor 5 (CCR5) that serves as an HIV receptor to entry in T-cells. More recently, similar approaches are also being applied to metabolic disorders. Visit the link ZFN clinical trials to learn more about their ongoing clinical trials. TALENs, which emerged soon after as a designer nuclease and energized the field by offering efficient and selective scissors for DNA-cutting, were followed by Cellectis in the strategy to generate allogeneic healthy donor T-cells for immunotherapies. Continuing with the Gardner's hype–hope cycle, the peak of expectation for gene editing has been tremendously accelerated by the advent of RNA-guided genome editing (RGEN) and the promise that CRISPR/Cas will give birth to medicines for multiple diseases. Indeed, we are taking part in the first seven clinical trials for anticancer therapy by knocking out the key PD-1 gene involved in immune-recognition and the elimination of tumor cells in China (CRISPR clinical trials).

Who are the main players in this new market? The main companies driving CRISPR technology and future applications in health are summarized in Table 2, exemplifying how fast and how strongly backed business has been created around preclinical companies based on this very recently developed technology. This high expectation is providing the Gartner cycle with a more pronounced slope than classical profiling.[7] Indeed, the three first publicly traded CRISPR-based biotech companies, Editas, Intellia Therapeutics, and CRISPR Therapeutics, had an initial public offering (IPO) price of around $100 Million each,[17] not 2 years after being found (Fig. 3).

5. A FUTURE SCENARIO—RISKS ASSOCIATED WITH GENE EDITING BUSINESSES

Genome-editing technologies are being envisaged as the Holy Grail for next generation precision medicine. However, these are targeted editing tools, and their precision, and specifically the degree to which there are off-target actions, and the clinical implications of such actions, are essential questions for the field. Other chapters in this volume have looked at the nature of the unwanted side effects of these gene-editing tools, and the techniques that may best serve as screening and comparison tools between platforms to reveal such errors introduced by editing.

In addition, the gene-editing business is subject to several risks that can be seen for an initial public offering of preclinical companies, among which the following stand out:

1. *Unpredictability*: Companies intend to identify and develop product candidates based on a new genome editing technology, which makes it difficult to predict the time and cost for product candidate development. No products that utilize genome editing technology have as yet been approved in the United States or in Europe, and there have only been a limited number of human-clinical trials of a genome-editing product candidate[14], of which only one trial has involved CRISPR/Cas9 technology.
2. *Short-lived*: The short operating history may make it difficult for investors to evaluate the success of the business to date, and to assess future viability.
3. *Regulatory issues*: Because genome editing is new and the regulatory landscape that will govern any product candidates is uncertain and may change, it cannot predict the time and cost of obtaining regulatory approval.

Table 2 Principal Companies Developing Strategies to use Therapeutic Gene Edition Tools[a]

Company	Founder	First round funding	Comments
Caribou Biosciences (Berkeley CA; USA)	Jennifer A Doudna	Atlas Venture $11 Million founder round (not went public)	• Found by scientists from the University of California, Berkeley to drive the commercialization of applications based on the remarkable nucleic acid modification capabilities found in prokaryotic CRISPR systems leaded by the pioneer scientist in CRIPSR technology JA Doudna
Intellia Therapeutics (Cambridge MA; USA) NASDAQ: NTLA	Caribou Biosciences and Atlas Venture	Atlas Venture and Novartis Institutes for Biomedical Research (NIBR)	• Novartis has exclusive rights to use Intellia's CRISPR platforms to develop CAR-T therapies, and nonexclusive rights for limited in vivo therapeutic applications
CRISPR Therapeutics (Basel; Switzerland) NASDAQ:CRSP	Emmanuelle Charpentier	Versant $25 Million first round	• Found by one of the coinventors of the CRISPR/cas9 technology • The company is focused on ex vivo programs involving gene editing of hematopoietic cells; in vivo programs targeting the liver and other organ systems, such as muscle and lung
Editas Medicine (Cambridge MA; USA) NASDAQ: EDIT	Feng Zhang; Jeniffer Doudna; George Church; Keith Young; David Liu	Founded by five of the world's leaders in gene editing $43 Million first round	• Editas has exclusive rights to the one issued patent for CRISPR granted to the Broad Institute and Harvard University (see text for details) • The company has nailed down key exclusive IP agreements with Massachusetts General Hospital (Boston MA) and Duke University (Durham, NC)

Sangamo Therapeutics (Richmond CA: USA) NASDAQ: SGMO	Edward O Lanphier II	Undisclosed	• Gene editing system based on zinc-finger nucleases • Sangamo has entered Phase 1/2 clinical trials as a possible functional cure for HIV and other malignancies (see text)
Cellectis (Paris, France) EPA:ALCLS	André Choulika	Undisclosed	• André Choulika is the inventor of nuclease-based genome editing and a pioneer in the analysis and use of meganucleases to modify complex genomes • Gene editing and cancer immunotherapy • Worldwide rights to a patent family titled "*Engineering Plant Genomes Using CRISPR/Cas Systems*"

[a]Data are mainly from the companies' websites and from www.nanalyze.com.

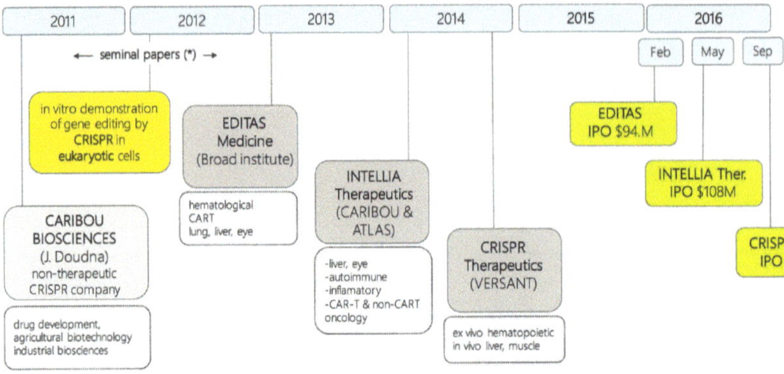

Fig. 3 The rapid development of the four main companies based on CRISPR/Cas technology, and their main therapeutic areas of interest for the introduction of gene editing products into clinics. Strategic alliances and more detailed information can be obtained from the companies' websites (links in Table 2).

4. *Social/demand*: The social perception of the genome-based medicines and therapeutic gene-editing based molecules may impact on demand for the potential products in accordance with cultural, religious and/or ethical issues, lowering, or even preventing access to the market.
5. *Unwanted side effects*: As happened before with CGT in the early 2000's, if serious adverse events, undesirable side effects, or unexpected characteristics are identified during the development of any gene editing product candidate, the technology will suffer a moratorium or even the withdrawal of the gene-medicine, or limits will be placed on the further clinical development of those product candidates. This issue can be affected by either the long-term follow-up of CGT clinical trials or from forthcoming genome editing clinical trials.
6. *Competitor/alternative technologies*: The genome-editing field is relatively new, and is evolving rapidly. Unforeseeable new techniques that may be discovered/engineered with unique properties, providing significant advantages over currently available genome editing technologies, may displace the existing ones.
7. *Regulatory endpoints*: Gene editing-based medicines are proof-of-concept for the treatment of diseases for which there is little clinical experience using new technologies. This increases the risk of delayed authorization from the regulatory authorities due to difficulties with evaluation or no sufficiently meaningful outcomes from the clinical trials, which lead to rapid approval.

8. *Intellectual property (IPR)*: As happened before with many other breakthrough-tagged products/technologies, such as enzyme replacement, monoclonal antibodies, phage display, or RNAi, licensed patents and patent applications are subject to priority disputes that might delay the commercialization of the medicinal product/technique. To follow the IP battle regarding the CRISPR system, there are excellent recent reviews.[18] Last February's decision by the Patent Trial and Appeal Board (PTAB) sided with the Broad Institute of MIT and Harvard, by finding "no interference in fact" between 12 patents related to CRISPR technology which list Broad's Feng Zhang, PhD as inventor and a patent application by Emmanuelle Charpentier, PhD and Jennifer Doudna, PhD, from UC Berkeley, concluded that "while the claims overlap, the scope of the claims made by UC and Broad did not define the same patentable invention." However, the decision did not determine who invented the use of the CRISPR/Cas9 genome-editing technology in eukaryotic cells. The UC Berkeley application states claims covering the use of CRISPR for a bacterial system, while the Broad Institute's patents focus on the use of CRISPR in eukaryotic systems, such as plants and higher animals. UC Berkeley contended that the application of CRISPR to eukaryotic systems represented an obvious rather than an inventive use and was thus nonpatentable. The battle is heading for further litigation, and UC Berkeley, U Vienna and E Charpentier have recently appealed to the US Court of Appeals for the Federal Circuit to overturn the February 15th decision.[19] The story is ongoing.

6. CONCLUSIONS

There is no doubt, that we are witnessing the advent of a revolutionary technology, with genome-editing aimed at changing the practice of the medicine in the near future, moving the practice from mostly the treatment of the symptoms of a disease toward an individualized precision medicine directed at the genetic basis of the disease. The new conception of medicine is in part focus on the genes that represent both medicines themselves, as in gene replacement therapy, or as entities susceptible to repair through editing.

REFERENCES

1. Woolf TM, Gurumurthy CB, Boyce F, Kmiec EB. To cleave or not to cleave: therapeutic gene editing with and without programmable nucleases. *Nat Rev Drug Discov.* 2017;16:296.
2. Kim JS. Genome editing comes of age. *Nat Protocols.* 2016;11:1573–1577.
3. Tebas P, et al. Gene editing of CCR5 in autologous CD4 T cells of persons infected with HIV. *N Engl J Med.* 2014;370:901–910.
4. Karpinski J, et al. Directed evolution of a recombinase that excises the provirus of most HIV-1 primary isolates with high specificity. *Nat Biotechnol* 2016;34:401–409
5. Lander ES. Brave New Genome. *N Engl J Med.* 2015;373:5–8.
6. Booth B. Cell and gene therapy: an "uutside-in" technology evolution. *LifeSciVC.* 2015 https://lifescivc.com/2015/11/cell-and-gene-therapy-an-outside-in-technology-evolution/.
7. Gardner J. Standing on the shoulders of stem cell gene therapists. *MedCitynews.* 2016 http://medcitynews.com/2016/07/stem-cell-gene-therapists/.
8. Check EA. Tragic setback. *Nature.* 2002;420:116–118.
9. Wilson JM. A history lesson for stem cells. *Science.* 2009;324:727–728.
10. Naldini L. Gene therapy returns to central stage. *Nature.* 2015;526:351–360.
11. Biffi A, et al. Lentiviral hematopoietic stem cell gene therapy benefits metachromatic leukodystrophy. *Science.* 2016;341:1233158–1233168.
12. Naldini L, Trono D, Verma IM. Lentiviral vectors, two decades later. *Science.* 2016;353:1101–1102.
13. European medicine Agency (EMA), UE. 2016. Available from: http://www.ema.europa.eu/docs/en_GB/document_library/EPAR_-_Summary_for_the_public/human/003854/WC500208202.pdf
14. Cyranosky D. CRISPR gene-editing tested in a person for the first time. *Nature.* 2016;539:479.
15. Corrigan-Curay J, et al. Genome editing technologies: defining a path to clinic. *Mol Ther.* 2015;23:796–806.
16. Fellman C, Gowe BG, Lin PC, Doudna JA, Corn JE. Cornerstones of CRIPSR-Cas in drug discovery and therapy. *Nat Rev Drug Discov.* 2017;16:89–101.
17. Adams B. CRISPR cuts out $90 M IPO bid as another gene editing biotech goes public. *FierceBiotech.* 2016 http://www.fiercebiotech.com/biotech/crispr-cuts-out-90m-ipo-bid-as-another-gene-editing-biotech-goes-public.
18. Sherkow JS. Law, history and lessons in the CRISPR patent conflict. *Nat Biotechnol.* 2015;33:256–257.
19. GEN NEW INSIGTHS: Judges Side with Broad Institute in CRISPR Patent Dispute. *Genetic Engineering & Biotechnologies News* 2017. http://www.genengnews.com/gen-news-highlights/judges-side-with-broad-institute-in-crispr-patent-dispute/81253890

INDEX

H

Q

R

S

FSC
www.fsc.org
MIX
Papier | Fördert
gute Waldnutzung
FSC® C083411

Kolibri 360 GmbH
Ferdinand-Jühlke-Straße 7
99095 Erfurt, Deutschland
produktsicherheit@kolibri360.de